Jae-Hyun Yang

Heisenberg Groups, Theta Functions and the Weil Representation

KYUNG MOON SA

2010 Mathematics Subject Classification. Primary 22E27, 14K25, 11F27.
keywords and phrases: Heisenberg group, theta functions, Weil representation.
This work was supported by Basic Science Program through the National Research Foundation of Korea(NRF) funded by the Ministry of Education, Science and Technology (41493-01).

Contents

Preface

A certain nilpotent Lie group plays an important role in the study of the foundations of quantum mechanics (cf. [30] and [41]) and the study of theta functions (see [4], [5], [14], [27], [28], [31], [39], [42], [43], [44] and [46]).

For any positive integers m and n, we consider the Heisenberg group

$$H_{\mathbb{R}}^{(n,m)} := \left\{ (\lambda, \mu, \kappa) \mid \lambda, \mu \in \mathbb{R}^{(m,n)}, \ \kappa \in \mathbb{R}^{(m,m)}, \ \kappa + \mu\,{}^t\lambda \text{ symmetric} \right\}$$

endowed with the following multiplication law

$$(\lambda, \mu, \kappa) \circ (\lambda', \mu', \kappa') = (\lambda + \lambda', \mu + \mu', \kappa + \kappa' + \lambda\,{}^t\mu' - \mu\,{}^t\lambda').$$

We let

$$H_{\mathbb{Z}}^{(n,m)} = \left\{ (\lambda, \mu, \kappa) \in H_{\mathbb{R}}^{(n,m)} \mid \lambda, \mu \in \mathbb{Z}^{(m,n)}, \ \kappa \in \mathbb{Z}^{(m,m)} \right\}$$

be the discrete subgroup of $H_{\mathbb{R}}^{(n,m)}$. The Heisenberg group $H_{\mathbb{R}}^{(n,m)}$ is embedded in the symplectic group $Sp(m+n, \mathbb{R})$ via the mapping

$$H_{\mathbb{R}}^{(n,m)} \ni (\lambda, \mu, \kappa) \longmapsto \begin{pmatrix} I_n & 0 & 0 & {}^t\mu \\ \lambda & I_m & \mu & \kappa \\ 0 & 0 & I_n & -{}^t\lambda \\ 0 & 0 & 0 & I_m \end{pmatrix} \in Sp(m+n, \mathbb{R}).$$

This Heisenberg group is a 2-step nilpotent Lie group and is important in the study of smooth compactification of the Siegel modular variety. In fact, $H_{\mathbb{R}}^{(n,m)}$ is obtained as the unipotent radical of the parabolic subgroup of the rational boundary component F_n(cf. [6] pp. 122-123, [29] p. 21 or [52] p. 36). In the case $m = 1$, the study on this Heisenberg group was done by many mathematicians, e.g., P. Cartier [4], J. Igusa [14], D. Mumford [27], [28] and many analysts(cf. [2]) explicitly. For the case $m > 1$, the multiplication law is a little different from that of the Heisenberg group which is usually known and needs much more complicated computation than the case $m = 1$.

The aim of this book is to investigate the Heisenberg group $H_{\mathbb{R}}^{(n,m)}$ in more detail and understand its relation to theta functions and the Weil representation. In the previous papers [42], [43] and [44], the author decomposed the L^2-space $L^2\left(H_{\mathbb{Z}}^{(n,m)}\backslash H_{\mathbb{R}}^{(n,m)}\right)$ with respect to the right regular representation of $H_{\mathbb{R}}^{(n,m)}$ explicitly and related the study of $H_{\mathbb{R}}^{(n,m)}$ to that of theta functions. We need to investigate $H_{\mathbb{R}}^{(n,m)}$ for the study of Jacobi forms (cf. [52], [58]), the study of harmonic analysis on the Jacobi group and degeneration of abelian varieties (cf. [6].

This book is organized as follows. The book consists of two chapters. In Section 1.1, we introduce the Heisenberg group $H_{\mathbb{R}}^{(n,m)}$ which will be investigated in the subsequent sections. And we find the Lie algebra of $H_{\mathbb{R}}^{(n,m)}$ and obtain the commutation relation for $H_{\mathbb{R}}^{(n,m)}$. In Section 1.2, we give an explicit description of theta functions due to J. Igusa (cf. [14] or [27]) and identify the theta functions with the smooth functions on $H_{\mathbb{R}}^{(n,m)}$ satisfying some conditions. The results of this section will be used later. In Section 1.3, using the Mackey decomposition of a locally compact group (cf. [24]), we introduce the induced representations of $H_{\mathbb{R}}^{(n,m)}$ and compute the unitary dual of $H_{\mathbb{R}}^{(n,m)}$. In Section 1.4, we realize the Schrödinger representation of $H_{\mathbb{R}}^{(n,m)}$ as the the representation of $H_{\mathbb{R}}^{(n,m)}$ induced by the one-dimensional unitary character of a certain subgroup of $H_{\mathbb{R}}^{(n,m)}$. In Section 1.5, we consider the Fock representation $\left(U^{F,\mathcal{M}}, \mathcal{H}_{F,\mathcal{M}}\right)$ of $H_{\mathbb{R}}^{(n,m)}$. We prove that for a positive definite symmetric half-integral matrix $\mathcal{M}$ of degree m, $U^{F,\mathcal{M}}$ is unitarily equivalent to the Schrödinger representation $U^{S,\mathcal{M}}$. We also find an orthonormal basis for the representation space $\mathcal{H}_{F,\mathcal{M}}$. This section is mainly based on the papers [31, 32, 45]. In Section 1.6, we prove that for any positive definite symmetric, half-integral matrix of degree m, the lattice representation $\pi_{\mathcal{M}}$ of $H_{\mathbb{R}}^{(n,m)}$ is unitarily equivalent to the $(\det 2\mathcal{M})^n$-multiples of the Schrödinger representation $U^{S,\mathcal{M}}$. We give a relation between the lattice representation $\pi_{\mathcal{M}}$ and theta functions. This section is based on the paper [46]. In Section 1.7, we find the coadjoint orbits of $H_{\mathbb{R}}^{(n,m)}$. And we describe explicitly the connection between the coadjoints orbits and the irreducible unitary representations of $H_{\mathbb{R}}^{(n,m)}$ following the work of A. Kirillov (cf. [16], [17] and [18]). In Section 1.8, considering the Schrödinger representation $\left(U^{S,I_m}, L^2(\mathbb{R}^{(m,n)}, d\xi)\right)$, we study the Hermite operators and the Hermite functions. We prove that Hermite functions defined in this section form an orthonormal basis for $L^2(\mathbb{R}^{(m,n)}, d\xi)$ and eigenfunctions for Hermite operators, the Fourier transform and the Fourier cotransform. We mention that

Hermitian functions are used to construct non-holomorphic modular forms of half-integral weight (cf. [43]). Implicitly the study of the Heisenberg group $H_{\mathbb{R}}^{(n,m)}$ implies that the confluent hypergeometric equations (in this case, the Hermite equations) are related to the study of automorphic forms. In Section 1.9, we investigate the irreducible components of $L^2\left(H_{\mathbb{Z}}^{(n,m)}\backslash H_{\mathbb{R}}^{(n,m)}\right)$. We describe the connection among these irreducible components, the Schrödinger representations, the Fock representations and the lattice representations explicitly. We also provide the orthonormal bases for these representation spaces respectively. A decomposition of $L^2(\Gamma\backslash G)$ for a general nilpotent Lie group G and a discrete subgroup Γ of G was investigated by C. C. Moore (cf. [26]).

In Section 2.1, we briefly review the symplectic group and its action on the Siegel upper half plane to be needed in the subsequent sections. We construct the universal covering group of the symplectic group. In Section 2.2, we present some properties of the geometry on the Siegel upper half plane which are used in the subsequent sections. In Section 2.3, we study the Weil representation associated to a positive definite symmetric real matrix of degree m. We describe the explicit actions for the Weil representation. We review the results on the Weil representation which were obtained by Kashiwara and Vergne [15]. In Section 2.4, we construct the covariant maps for the Weil representation. In Section 2.5, we review various type of theta series associated to quadratic forms. In Section 2.6, we discuss the theta series with harmonic coefficients. Pluriharmonic polynomials play an important role in the study of the Weil representation. We prove that the theta series with pluriharmonic polynomials as coefficients are a modular form for a suitable congruence subgroup of the Siegel modular group. This section is mainly based on the book [28] which dealt with the case $m = 1$. In Section 2.7, we investigate the relation between the Weil representation and the theta series. We construct modular forms using the covariant maps for the Weil representation. In Section 2.8, we discuss the spectral theory on the principally polarized abelian variety A_Ω attached to an element Ω of the Siegel upper half plane. We decompose the L^2-space of A_Ω into irreducibles explicitly. We refer to [47] for more detail.

I want to mention that a Heisenberg group was paid to an attention by some differential geometers, e.g., M. L. Gromov, in the sense of a parabolic geometry. A Heisenberg group is regarded as a principal fibre bundle over an Euclidean space with a vector space or a circle as fibres and may be also regarded as the boundary of a complex ball. The geometry of this group is quite different from that of an Euclidean space. We refer to the interesting paper [5] of P. Deligne.

Professor Dr. Friedrich Hirzebruch passed away on May 27, 2012. I would like to send my condolences for his death. He was very kind to me during my stay at MPIM, Bonn. I dedicate this book to him. Finally I hope that this book will be very useful in the study of the theory of harmonic analysis on the Siegel-Jacobi space.

Acknowledgements: The author thanks the National Research Foundation of Korea for its financial support. This work was in part done during his stay at the Max-Planck-Institut für Mathematik in Bonn. He is also grateful to MPIM, Bonn for its hospitality.

Notations: We denote by $\mathbb{Z}$, $\mathbb{R}$ and $\mathbb{C}$ the ring of integers, the field of real numbers, and the field of complex numbers respectively. $\mathbb{C}^*$ denotes the multiplicative group consisting of all nonzero complex numbers. $\mathbb{C}_1^*$ denotes the multiplicative group consisting of all complex numbers z with $|z| = 1$. $Sp(n, \mathbb{R})$ denotes the symplectic group of degree n. $\mathbb{H}_n$ denotes the Siegel upper half plane of degree n. The symbol ":=" means that the expression on the right is the definition of that on the left. We denote by $\mathbb{Z}^+$ the set of all positive integers. $F^{(k,l)}$ denotes the set of all $k \times l$ matrices with entries in a commutative ring F. For any $M \in F^{(k,l)}$, ${}^tM$ denotes the transpose of a matrix M. For a complex matrix A, $\overline{A}$ denotes the complex *conjugate* of A. The diagonal matrix with entries $a_1, \cdots, a_n$ on the diagonal position is denoted by $\mathrm{diag}(a_1, \cdots, a_n)$. For $A \in F^{(k,k)}$, $\sigma(A)$ denotes the trace of A. For $A \in F^{(k,l)}$ and $B \in F^{(k,k)}$, we set $B[A] = {}^tABA$. I_k denotes the identity matrix of degree k. For a positive integer m, $\mathrm{Sym}\,(m, K)$ denotes the vector space consisting of all symmetric $m \times m$ matrices with entries in a commutative ring K. If H is a complex matrix or a complex bilinear form on a complex vector space, $\mathrm{Re}\, H$ and $\mathrm{Im}\, H$ denote the real part of H and the imaginary part of H respectively. If X is a space, $\mathcal{S}(X)$, $C(X)$ and $C_c^\infty(X)$ denotes the Schwarz space of infinitely differentiable functions on X that are rapidly decreasing at infinity, the space of all continuous functions on X and the vector space consisting of all compactly supported and infinitely differentiable functions on X respectively. We inroduce the following notations:

$$\begin{aligned}
\mathbb{Z}_{\geq 0}^{(m,n)} &= \left\{ J = (J_{ka}) \in \mathbb{Z}^{(m,n)} \mid J_{ka} \geq 0 \ \text{ for all } k, a \right\}, \\
|J| &= \sum_{k,a} J_{ka} \quad \text{ for } J = (J_{ka}) \in \mathbb{Z}_{\geq 0}^{(m,n)}, \\
J \pm \epsilon_{ka} &= (J_{11}, \cdots, J_{ka} \pm 1, \cdots, J_{mn}), \\
J! &= J_{11}! \cdots J_{ka}! \cdots J_{mn}!.
\end{aligned}$$

For $\xi = (\xi_{ka}) \in \mathbb{R}^{(m,n)}$ or $\mathbb{C}^{(m,n)}$ and $J = (J_{ka}) \in \mathbb{Z}_{\geq 0}^{(m,n)}$, we denote

$$\xi^J = \xi_{11}^{J_{11}} \xi_{12}^{J_{12}} \cdots \xi_{ka}^{J_{ka}} \cdots \xi_{mn}^{J_{mn}}.$$

Table of Symbols

Chapter 1 Heiseberg Groups

SECTION 1.1 : $\mathcal{A}$, $\widehat{\mathcal{A}}$, $\mathcal{S}$, α_λ, α_λ^*, $\widehat{\mathcal{O}}_{\hat{\kappa}}$, $\widehat{\mathcal{O}}_{\hat{y}}$, $S_{\hat{\kappa}}$, X_{kl}^0, X_{ka}, $\widehat{X}_{lb}$, D_{kl}^0, D_{ka}, $\widehat{D}_{lb}$, Z_{kl}^0, Y_{ka}^+, Y_{lb}^-, $E_{kl}^\bullet$, R_{kl}, P_{ka}, Q_{lb}

SECTION 1.2 : Ω, $R_{\mathcal{M}}^\Omega$, $\vartheta^{(S)}\begin{bmatrix}A\\B\end{bmatrix}(\Omega,W)$, $Q_\xi(W)$, $J(\xi,W)$, $l(\xi)$, $\mathrm{Her}\,Q$, $\psi(\xi)$, $L(Q,l,\psi)$, $\mathrm{Sym}\,Q$, $Th(H,\psi,L)$, $\chi_{S,\Omega,A,B}$, $q_{S,\Omega}$, $H_{S,\Omega}$, $\psi_{S,\Omega}$, L_Ω, $A_{S,\Omega}$, R_S^Ω, Θ, $Th(H_{S,\Omega},\ \psi_{S,\Omega}, L_\Omega)$, $R_{S,A,B}^\Omega$, $\Theta_{A,B}$, $J_{S,\Omega,A,B}$, $\tilde{J}_{S,\Omega,A,B}$, $\mathcal{A}_{S,\Omega}$, $\mathcal{L}_X$, φ_f

SECTION 1.3 : U_σ, $(\ ,\ )_{\mathcal{H}}$, $\mathcal{H}_\sigma$, k_g, s_g, $T_{\hat{\kappa}}$, $T_{\hat{x},\hat{y}}, \chi_{\hat{x}}$

SECTION 1.4 : G, K, k_g, s_g, U_{σ_c}, U_c, $\mathcal{H}^{\sigma_c}$, $\mathcal{H}^c$, $\mathcal{H}_{\sigma_c}$, $\mathcal{H}_c$, Φ_c, $dU_c(X)$, $f_{c,J}$

SECTION 1.5 : P_{ka}, Q_{lb}, $\mathbf{A}$, J, $J_{\mathbb{C}}$, V^+, V^-, T, $\mathbf{H}$, V_*, $G_{\mathbb{C}}$, z^0, z^1, R^+, R^-, z^+, z^-, $U^{F,c}$, $\mathcal{H}^{F,c}$, δ_c, $\mathcal{H}_{F,c}$, $(\ ,\ )_{F,c}$, Λ, Λ_f, Δ, Δ_ψ, $d\mu(W)$, $\mathcal{H}_{m,n}$, $\Phi_J(W), \kappa(W,W'), \|\ f\ \|_{\mathcal{M}}$, $(\ ,\ )_{\mathcal{M}}$, $d\mu_{\mathcal{M}}(W)$, $\Phi_{\mathcal{M},J}(W)$, $k(U,W)$, $\mathcal{I}(W,W')$, $k_{\mathcal{M}}(U,W)$, $I_{\mathcal{M}}(W,W')$, $U^{S,\mathcal{M}}$, $I_{\mathcal{M}}$, h_J, $A_{\mathcal{M}}(U,W)$, $dU^{F,\mathcal{M}}(X)$

SECTION 1.6 : L_B^*, Γ_L, $\Gamma_{L_B^*}$, $\mathcal{Z}_0$, $\phi_{k,l}$, $\phi_{\mathcal{M},q}$, $\pi_{\mathcal{M},q}$, $\mathcal{H}_{\mathcal{M}}$, $\mathcal{T}$, $\phi_{\mathcal{M},\alpha}$, $\mathcal{H}_{\mathcal{M},\alpha}$, $\vartheta_{\mathcal{M},\alpha}, q_{\mathcal{M}}$, $\varphi_{\mathcal{M},q_{\mathcal{M}}}$, $\mathcal{H}_{\mathcal{M},q_{\mathcal{M}}}$, $\pi_{\mathcal{M},q_{\mathcal{M}}}$, $\pi_{\mathcal{M},q_{\mathcal{M}}}$, $\mathbf{H}_{\mathcal{M},q_{\mathcal{M}}}$, E_ϕ, F_ϕ, $F_{\Omega,\phi}$, $\vartheta_{\Omega,\phi}$

SECTION 1.7 : $\mathfrak{g}$, $\mathfrak{g}^*$, $F(a,b,c)$, Ad_G^*, $\Omega_{a,b}$, Ω_c, $\mathcal{O}(G)$, $\widehat{G}$, B_F, $ad_{\mathfrak{g}}^*$, G_F, $\mathfrak{g}_F$, $\mathrm{rad}\,B_F$, Ω_F, $\widetilde{X}$, B_{Ω_F}, $\pi_{a,b}$, $\mathfrak{k}$, $\chi_{c,\mathfrak{k}}$, $\pi_{c,\mathfrak{k}}$, π_c, χ_c, π_c^1, $\mathcal{CF}_{\mathfrak{g}}$, $C_c^\infty(G)$, $C_c^\infty(\mathfrak{g})$, $C(\mathfrak{g}^*)$, $\mathcal{S}(G/\mathcal{Z})$, π_c^1, $L^2(G/Z,\chi_c)$, $TC\big(L^2(\mathbb{R}^{(m,n)},d\xi)\big)$, $HS\big(L^2(\mathbb{R}^{(m,n)},d\xi)\big)$, Ad_K^*, $\omega_{b,c}$, $\chi_{b,c}$, $p(\Omega_c)$

SECTION 1.8 : $dU_{I_m}(X)$, A_{ka}^+, A_{lb}^-, C_{kl}, f_0, f_J, h_J, H_{ka}, P_J, ∂_{ka}, $U(X)$, A^+, A^-, $c_{k,p}$, $d_{k,p}$, $b_{k,p}$

SECTION 1.9 : $d\xi_{\Omega,\mathcal{M}}$, $\mathcal{T}$, $\mathcal{L}$, $\Phi_J^{(\mathcal{M})}\begin{bmatrix}A_\alpha\\0\end{bmatrix}(\Omega|\cdot)$, Γ_G, $H_\Omega^{(\mathcal{M})}\begin{bmatrix}A_\alpha\\0\end{bmatrix}$, ρ, $f_{\Omega,J}^{(\mathcal{M})}$, $\Phi_{\Omega,\alpha}^{(\mathcal{M})}$, $\Delta_{\Omega,\mathcal{M}}$, $H_J(\xi)$, $\vartheta_{\mathcal{M},\alpha,J}$, $H_J^{(\mathcal{M})}\begin{bmatrix}A_\alpha\\0\end{bmatrix}(\Omega|\cdot)$, $R(c)$

Chapter 2 Theta Functions and the Weil Representation

SECTION 2.1 : $Sp(n,\mathbb{R})$, J_n, Γ_n, $(\Gamma_n)_\Omega$, $\Gamma_n(q)$, $\Gamma_{\vartheta,n}$, $\Gamma_{n,0}(q)$, Ω^*, X^*, Y^*, dM_Ω, $T_\Omega(\mathbb{H}_n)$, (V,B), $L^\perp$, $Sp(B)$, $\tau(L_1,L_2,L_3)$, Λ, $\widetilde{\Lambda}$, $\mathscr{U}$, $\tau(L_1,L_2,\cdots,L_k)$, $U(L_1,u_1;\mathscr{U},L_2)$, $\pi:\widetilde{\Lambda}\longrightarrow\Lambda$, L_*, $\widetilde{Sp(B)}_*$, $\mathscr{E}$, $\mathscr{W}(\mathscr{E},L_2)$, (V,ε), $\delta(A)$, $g_{M,L}$, L^+, $s_L(g)$, $\xi\big((L_1,\varepsilon_1),(L_2,\varepsilon_2)\big)$, $s\big((L_1,\varepsilon_1),(L_2,\varepsilon_2)\big)$, $\widetilde{s}_*$, $Sp(B)_*$, $c_*(g_1,g_2)$, $\varphi(g,n)$, $Mp(B)_*$

SECTION 2.2 : ds^2, dv_n, $R(\Omega_0,\Omega_1)$, $\rho(\Omega_0,\Omega_1)$, $\mathbb{D}_n$, Ψ, T, G_*, $SU(n,n)$, P^+, K_*, ds^2_*, Δ_*, K, $\mathfrak{sp}(n,\mathbb{R})$, $\mathfrak{k}$, $\mathfrak{p}$, ψ, δ, $\mathbb{T}_n$, $\mathrm{Pol}(\mathbb{T}_n)$, Φ, f_Z, Φ, $\mathcal{P}_n$, $\mathcal{R}_n$, $\partial\mathcal{R}_n$. $d\mu_n, \mathcal{F}_n$

SECTION 2.3 : U_c, G^J, U_c^M, R_c, $\alpha_c(M_1,M_2)$, $J(M,\Omega)$, $J^*(M,\Omega)$, $\widetilde{R}_c$, $s_c(M)$, $Sp(n,\mathbb{R})_*$, $Mp(n,\mathbb{R})$, ω_c, t_b, d_a, σ_n, $T_c(t_b)$, $O(m)$, $A_c(d_a)$, $B_c(\sigma_n)$, (σ,V_σ), $L^2\big(\mathbb{R}^{(m,n)};\sigma\big)$, $\omega_c(\sigma)$, $\widehat{O(m)}$, Σ_m, $K=U(n)$, $\widehat{K}$, $\mathcal{O}(\mathbb{H}_n,V_\tau)$, T_τ, Φ_τ, $\mathfrak{H}$, $\tau(\sigma)$, $\mathfrak{H}(\sigma)$, σ^*, $\mathscr{F}_\sigma$

SECTION 2.4 : $\mathscr{F}^{(c)}$, $J_m(M,\Omega)$

SECTION 2.5 : $A(S,T)$, S, $\vartheta_S(\Omega)$, $h(\Omega)$, $\vartheta_{S;A,B}(\Omega)$, $\vartheta(\Omega;a,b)$, $\{a,b\}$, $\gamma\diamond\begin{pmatrix}a\\b\end{pmatrix}$, $\mathscr{C}^e$, $\nu(\gamma)$, t_S, k_n, $\Delta^{(n)}(\Omega)$, $\nu_S(\gamma)$

SECTION 2.6 : $\vartheta_S\begin{bmatrix}\alpha\\\beta\end{bmatrix}(\Omega,Z)$, $\chi\begin{bmatrix}\alpha\\\beta\end{bmatrix}(N)$, $\mathfrak{P}_{m,n}$, $\vartheta_{S,P}\begin{bmatrix}\alpha\\\beta\end{bmatrix}(\Omega,Z)$, $\vartheta_{S,P}(\Omega,Z)$, $\vartheta_{S,P}\begin{bmatrix}\alpha\\\beta\end{bmatrix}(\Omega)$, $P(\partial)$, $\mathfrak{P}_N$, $\langle P,Q\rangle$, $\mathfrak{H}(S)$, I, h_{ij}, $T=(t_{kl})$, $\mathfrak{P}_{m,n}$, $\mathfrak{H}(S)_\mathbb{R}$, $I_\mathbb{R}$, $f_{A,B}$, $P_{A,B}$, $\mathfrak{H}(S)^\perp$, $O(S)$, $\widetilde{P}(Z)$, $\vartheta_W\begin{bmatrix}\alpha\\\beta\end{bmatrix}$, $\widetilde{GL(n,\mathbb{C})}$, $\mathcal{L}^{\frac{1}{2}}$, $\widetilde{M}$, $\mathcal{L}^{\frac{k}{2}}$, $\mathfrak{H}_{m,n}(\rho)$

SECTION 2.7 : (π,V_π), $\mathscr{F}$, θ, $\Theta(\Omega)$, $\Theta_\mathcal{M}(\Omega)$, ϑ, $\mathscr{F}^\mathcal{M}$, $\omega_\mathcal{M}$, $\widehat{f}$

SECTION 2.8 : $\mathbb{H}_n\times\mathbb{C}^{(m,n)}$, $\mathbb{H}_{n,m}$, $\Gamma_{n,m}$, E_{kj}, $F_{kj}(\Omega)$, L_Ω, $\Omega_\flat$, A_Ω, $\Delta_{n,m}$, $\mathcal{F}_{n,m}$, Δ_Ω, $L^2(A_\Omega)$, $\mathrm{Im}\,\Omega$, $E_{\Omega;A,B}(Z)$, ds^2_Ω, Γ^J, $H_\mathbb{Z}^{(n,m)}$, $||f||_\Omega$, dv_Ω, $(f,g)_\Omega$, $L^2(T)$, $E_{A,B}(W)$, Δ_T, Φ_Ω

Chapter 1
Heisenberg Groups

1.1 The Heisenberg Group $H_{\mathbb{R}}^{(n,m)}$

For any two positive integer m and n, we let

$$H_{\mathbb{R}}^{(n,m)} = \left\{ (\lambda, \mu, \kappa) \,\middle|\, \lambda, \mu \in \mathbb{R}^{(m,n)},\ \kappa \in \mathbb{R}^{(m,m)},\ \kappa + \mu\, {}^t\lambda \text{ symmetric} \right\}$$

be the Heisenberg group endowed with the following multiplication law

$$(\lambda, \mu, \kappa) \circ (\lambda_0, \mu_0, \kappa_0) := (\lambda + \lambda_0, \mu + \mu_0, \kappa + \kappa_0 + \lambda\, {}^t\mu_0 - \mu\, {}^t\lambda_0). \tag{1.1.1}$$

We observe that $H_{\mathbb{R}}^{(n,m)}$ is a 2-step nilpotent Lie group. It is easy to see that the inverse of an element $(\lambda, \mu, \kappa) \in H_{\mathbb{R}}^{(n,m)}$ is given by

$$(\lambda, \mu, \kappa)^{-1} = (-\lambda, -\mu, -\kappa + \lambda\, {}^t\mu - \mu\, {}^t\lambda).$$

Now we put

$$[\lambda, \mu, \kappa] = (0, \mu, \kappa) \circ (\lambda, 0, 0) = (\lambda, \mu, \kappa - \mu\, {}^t\lambda). \tag{1.1.2}$$

Then $H_{\mathbb{R}}^{(n,m)}$ may be regarded as a group equipped with the following multiplication

$$[\lambda, \mu, \kappa] \diamond [\lambda_0, \mu_0, \kappa_0] = [\lambda + \lambda_0, \mu + \mu_0, \kappa + \kappa_0 + \lambda\, {}^t\mu_0 + \mu_0\, {}^t\lambda]. \tag{1.1.3}$$

The inverse of $[\lambda, \mu, \kappa] \in H_{\mathbb{R}}^{(n,m)}$ is given by

$$[\lambda, \mu, \kappa]^{-1} = [-\lambda, -\mu, -\kappa + \lambda\, {}^t\mu + \mu\, {}^t\lambda].$$

We set

$$\mathcal{A} = \left\{ [0, \mu, \kappa] \in H_{\mathbb{R}}^{(n,m)} \,\middle|\, \mu \in \mathbb{R}^{(m,n)},\ \kappa = {}^t\kappa \in \mathbb{R}^{(m,m)} \right\}. \tag{1.1.4}$$

Then $\mathcal{A}$ is a commutative normal subgroup of $H_{\mathbb{R}}^{(n,m)}$. Let $\widehat{\mathcal{A}}$ be the Pontrajagin dual of $\mathcal{A}$, i.e., the commutative group consisting of all unitary characters of $\mathcal{A}$. Then $\widehat{\mathcal{A}}$ is isomorphic to the additive group $\mathbb{R}^{(m,n)} \times \mathrm{Sym}\,(m,\mathbb{R})$ via

$$\langle a, \hat{a} \rangle := e^{2\pi i \sigma(\hat{\mu}\,{}^t\mu + \hat{\kappa}\kappa)}, \quad a = [0,\mu,\kappa] \in \mathcal{A},\ \hat{a} = (\hat{\mu},\hat{\kappa}) \in \widehat{\mathcal{A}}. \tag{1.1.5}$$

We put

$$\mathcal{S} = \left\{ [\lambda,0,0] \in H_{\mathbb{R}}^{(n,m)} \,\middle|\, \lambda \in \mathbb{R}^{(m,n)} \right\} \cong \mathbb{R}^{(m,n)}. \tag{1.1.6}$$

Then $\mathcal{S}$ acts on $\mathcal{A}$ as follows:

$$\alpha_\lambda([0,\mu,\kappa]) := [0,\mu,\kappa + \lambda\,{}^t\mu + \mu\,{}^t\lambda], \quad \alpha_\lambda = [\lambda,0,0] \in \mathcal{S}. \tag{1.1.7}$$

It is easy to see that the Heisenberg group $\left(H_{\mathbb{R}}^{(n,m)}, \diamond\right)$ is isomorphic to the semidirect product $G_H := \mathcal{S} \ltimes \mathcal{A}$ of $\mathcal{A}$ and $\mathcal{S}$ whose multiplication is given by

$$(\lambda, a)\cdot(\lambda_0, a_0) = (\lambda + \lambda_0, a + \alpha_\lambda(a_0)), \quad \lambda,\lambda_0 \in \mathcal{S},\ a, a_0 \in \mathcal{A}.$$

On the other hand, $\mathcal{S}$ acts on $\widehat{\mathcal{A}}$ by

$$\alpha_\lambda^*(\hat{a}) := (\hat{\mu} + 2\hat{\kappa}\lambda, \hat{\kappa}), \quad [\lambda,0,0] \in \mathcal{S},\ \hat{a} = (\hat{\mu},\hat{\kappa}) \in \widehat{\mathcal{A}}. \tag{1.1.8}$$

Then we have the relation $\langle \alpha_\lambda(a), \hat{a} \rangle = \langle a, \alpha_\lambda^*(\hat{a}) \rangle$ for all $a \in \mathcal{A}$ and $\hat{a} \in \widehat{\mathcal{A}}$.

We have two types of $\mathcal{S}$-orbits in $\widehat{\mathcal{A}}$.

Type I. Let $\hat{\kappa} \in \mathrm{Sym}\,(m,\mathbb{R})$ with $\hat{\kappa} \neq 0$. The $\mathcal{S}$-orbit of $\hat{a}(\hat{\kappa}) := (0,\hat{\kappa}) \in \widehat{\mathcal{A}}$ is given by

$$\widehat{\mathcal{O}}_{\hat{\kappa}} := \left\{ (2\hat{\kappa}\lambda, \hat{\kappa}) \in \widehat{\mathcal{A}} \,\middle|\, \lambda \in \mathbb{R}^{(m,n)} \right\} \cong \mathbb{R}^{(m,n)}. \tag{1.1.9}$$

Type II. Let $\hat{y} \in \mathbb{R}^{(m,n)}$. The $\mathcal{S}$-orbit $\widehat{\mathcal{O}}_{\hat{y}}$ of $\hat{a}(\hat{y}) := (\hat{y}, 0)$ is given by

$$\widehat{\mathcal{O}}_{\hat{y}} := \{\, (\hat{y},0) \,\} = \{\hat{a}(\hat{y})\}. \tag{1.1.10}$$

We have

$$\widehat{\mathcal{A}} = \left(\bigcup_{\hat{\kappa}\in \mathrm{Sym}(m,\mathbb{R})} \widehat{\mathcal{O}}_{\hat{\kappa}} \right) \bigcup \left(\bigcup_{\hat{y}\in\mathbb{R}^{(m,n)}} \widehat{\mathcal{O}}_{\hat{y}} \right)$$

as a set. The stabilizer $\mathcal{S}_{\hat{\kappa}}$ of $\mathcal{S}$ at $\hat{a}(\hat{\kappa}) = (0,\hat{\kappa})$ is given by

$$\mathcal{S}_{\hat{\kappa}} = \{0\}. \tag{1.1.11}$$

And the stabilizer $\mathcal{S}_{\hat{y}}$ of $\mathcal{S}$ at $\hat{a}(\hat{y}) = (\hat{y},0)$ is given by

$$\mathcal{S}_{\hat{y}} = \left\{ [\lambda,0,0] \,\middle|\, \lambda \in \mathbb{R}^{(m,n)} \right\} = \mathcal{S} \cong \mathbb{R}^{(m,n)}. \tag{1.1.12}$$

The following matrices

$$X_{kl}^0 := \begin{pmatrix} 0\,0\,0 & 0 \\ 0\,0\,0 & \frac{1}{2}(E_{kl}+E_{lk}) \\ 0\,0\,0 & 0 \\ 0\,0\,0 & 0 \end{pmatrix}, \qquad 1 \le k \le l \le m,$$

$$X_{ka} := \begin{pmatrix} 0 & 0\,0 & 0 \\ E_{ka} & 0\,0 & 0 \\ 0 & 0\,0 & -{}^tE_{ka} \\ 0 & 0\,0 & 0 \end{pmatrix}, \qquad 1 \le k \le m,\ 1 \le a \le n,$$

$$\widehat{X}_{lb} := \begin{pmatrix} 0\,0 & 0 & {}^tE_{lb} \\ 0\,0 & E_{lb} & 0 \\ 0\,0 & 0 & 0 \\ 0\,0 & 0 & 0 \end{pmatrix}, \qquad 1 \le l \le m,\ 1 \le b \le n$$

form a basis of the Lie algebra $\mathcal{H}_{\mathbb{R}}^{(n,m)}$ of the real Heisenberg group $H_{\mathbb{R}}^{(n,m)}$. Here E_{kl} denotes the $m \times m$ matrix with entry 1 where the k-th row and the l-th column meet, all other entries 0 and E_{ka} (resp. E_{lb}) denotes the $m \times n$ matrix with entry 1 where the k-th (resp. the l-th) row and the a-th (resp. the b-th) column meet, all other entires 0. By an easy calculation, we see that the following vector fields

$$D_{kl}^0 := \frac{\partial}{\partial \kappa_{kl}}, \qquad 1 \le k \le m,$$

$$D_{ka} := \frac{\partial}{\partial \lambda_{ka}} - \left(\sum_{p=1}^{k} \mu_{pa} \frac{\partial}{\partial \kappa_{pk}} + \sum_{p=k+1}^{m} \mu_{pa} \frac{\partial}{\partial \kappa_{kp}} \right), \quad 1 \le k \le m,\ 1 \le a \le n,$$

$$\widehat{D}_{lb} := \frac{\partial}{\partial \mu_{lb}} + \left(\sum_{p=1}^{l} \lambda_{pb} \frac{\partial}{\partial \kappa_{pl}} + \sum_{p=l+1}^{m} \lambda_{pb} \frac{\partial}{\partial \kappa_{lp}} \right), \quad 1 \le k \le m,\ 1 \le a \le n$$

form a basis for the Lie algebra of left-invariant vector fields on the Lie group $H_{\mathbb{R}}^{(n,m)}$.

Lemma 1. *We have the following Heisenberg commutation relations*

$$\begin{aligned} [\,D_{kl}^0, D_{st}^0\,] &= [\,D_{kl}^0, D_{sa}\,] = [\,D_{kl}^0, \widehat{D}_{sa}\,] = 0, \\ [\,D_{ka}, D_{lb}\,] &= [\,\widehat{D}_{ka}, \widehat{D}_{lb}\,] = 0, \\ [\,D_{ka}, \widehat{D}_{lb}\,] &= 2\,\delta_{ab}\, D_{kl}^0, \end{aligned}$$

where $1 \le k, l, s, t \le m$, $1 \le a, b \le n$ *and* δ_{ab} *denotes the Kronecker delta symbol.*

Proof. The proof follows from a straightforward calculation. □

We put

$$\begin{aligned} Z_{kl}^0 &:= -\sqrt{-1}\, D_{kl}^0, \quad 1 \le k \le l \le m, \\ Y_{ka}^+ &:= \frac{1}{2}\,(D_{ka} + \sqrt{-1}\,\widehat{D}_{ka}), \quad 1 \le k \le m,\ 1 \le a \le n, \\ Y_{lb}^- &:= \frac{1}{2}\,(D_{lb} - \sqrt{-1}\,\hat{D}_{lb}), \quad 1 \le l \le m,\ 1 \le b \le n. \end{aligned}$$

Then it is easy to see that the vector fields Z_{kl}^0, Y_{ka}^+, Y_{lb}^- form a basis of the complexification of the real Lie algebra $\mathcal{H}_{\mathbb{R}}^{(n,m)}$.

Lemma 2. *We have the following commutation relations*

$$\begin{aligned} &[Z_{kl}^0, Z_{st}^0] = [Z_{kl}^0, Y_{sa}^+] = [Z_{kl}^0, Y_{sa}^-] = 0, \\ &[Y_{ka}^+, Y_{lb}^+] = [Y_{ka}^-, Y_{lb}^-] = 0, \\ &[Y_{ka}^+, Y_{lb}^-] = \delta_{ab}\, Z_{kl}^0, \end{aligned}$$

where $1 \le k, l, s, t \le m$ *and* $1 \le a, b \le n$.

Proof. It follows immediately from Lemma 1. □

We let $E_{kl}^\bullet := E_{kl} + E_{lk}$ for $1 \le k \le l \le m$. We put

$$\begin{aligned} R_{kl}(r) &:= \exp\left(2rX_{kl}^0\right) = (0,0,rE_{kl}^\bullet), \quad r \in \mathbb{R}, \\ P_{sa}(x) &:= \exp\left(xX_{sa}\right) = (xE_{sa},0,0), \quad x \in \mathbb{R}, \\ Q_{tb}(y) &:= \exp\left(y\widehat{X}_{tb}\right) = (0,yE_{tb},0), \quad y \in \mathbb{R}, \end{aligned}$$

where $1 \le k \le l \le m$, $1 \le s, t \le m$ and $1 \le a, b \le n$. Then these one-parameter subgroups generate the Heisenberg group $H_{\mathbb{R}}^{(n,m)}$. They satisfy the **Weyl commutation relations**:

$$P_{sa}(x) \circ Q_{sa}(y) = Q_{sa}(y) \circ P_{sa}(x) \circ R_{ss}(xy) \quad (\text{all others commute}),$$

where $1 \le s \le m$ and $1 \le a \le n$.

J. von Neumann [30] and M. Stone [38] proved the following uniqueness theorem simultaneously and independently.

Theorem 1. *Let* π_1 *and* π_2 *be two irreducible unitary representations of the Heisenberg group* $H_{\mathbb{R}}^{(n,m)}$ *such that*

$$\pi_1((0,0,\kappa)) = \pi_2((0,0,\kappa)) \qquad \text{for all } \kappa = {}^t\kappa \in \mathbb{R}^{(m,m)}.$$

Then π_1 *is unitarily equivalent to* π_2.

We omit the proof of the above theorem. We refer to [21] for the proof of Theorem 1 in the case $m = 1$ and also to [4] and [28] for more detail.

1.2 Theta Functions

We fix an element $\Omega \in \mathbb{H}_n$ once and for all. From now on, we put $i = \sqrt{-1}$. Let $\mathcal{M}$ be a positive definite, symmetric even integral matrix of degree m. A holomorphic function $f : \mathbb{C}^{(m,n)} \longrightarrow \mathbb{C}$ satisfying the following equation

$$f(W + \xi\Omega + \eta) = e^{-\pi i\,\sigma\{\mathcal{M}(\xi\Omega\,{}^t\xi + 2\,\xi\,{}^tW)\}} f(W), \quad W \in \mathbb{C}^{(m,n)} \tag{1.2.1}$$

for all $\xi,\, \eta \in \mathbb{Z}^{(m,n)}$ is called a **theta function of level** $\mathcal{M}$ with respect to Ω. The set $R^{\Omega}_{\mathcal{M}}$ of all theta functions of level $\mathcal{M}$ with respect to Ω is a complex vector space of dimension $(\det\, \mathcal{M})^n$ with a basis consisting of theta functions

$$\vartheta^{(\mathcal{M})}\begin{bmatrix} A \\ 0 \end{bmatrix}(\Omega, W) := \sum_{N \in \mathbb{Z}^{(m,n)}} e^{\pi i\,\sigma\{\mathcal{M}((N+A)\Omega\,{}^t(N+A) + 2\,W\,{}^t(N+A))\}}, \tag{1.2.2}$$

where A runs over a complete system of the cosets $\mathcal{M}^{-1}\mathbb{Z}^{(m,n)}/\mathbb{Z}^{(m,n)}$.

Definition 1. Let S be a positive definite, symmetric real matrix of degree m and let $A, B \in \mathbb{R}^{(m,n)}$. We define the theta function

$$\vartheta^{(S)}\begin{bmatrix} A \\ B \end{bmatrix}(\Omega, W) = \sum_{N \in \mathbb{Z}^{(m,n)}} e^{\pi i\,\sigma\{S((N+A)\Omega\,{}^t(N+A) + 2\,(W+B)\,{}^t(N+A))\}} \tag{1.2.3}$$

with characteristic (A, B) converging normally on $\mathbb{H}_n \times \mathbb{C}^{(m,n)}$.

We have a general definition of theta functions.

Definition 2. Let V be a complex vector space and let $L \subset V$ be a lattice of V. A **theta function on** V **relative to** L is a nonzero holomorphic function ϑ on V satisfying the following condition

$$\vartheta(W + \xi) = e^{2\pi i(Q_\xi(W) + c_\xi)}\vartheta(W),$$

where Q_ξ is a $\mathbb{C}$-linear form on V and c_ξ is an element of $\mathbb{C}$, for every $W \in V$ and $\xi \in L$.

If ϑ is a theta function on V relative to L, then the mapping $J_\vartheta : L \times V \longrightarrow \mathbb{C}^*$ defined by

$$J_\vartheta(\xi, W) := e^{2\pi i(Q_\xi(W) + c_\xi)}, \quad \xi \in L, \quad W \in V$$

is easily seen to be an automorphic factor. This means that J_ϑ satisfies the following condition

$$J_\vartheta(\xi_1 + \xi_2, W) = J_\vartheta(\xi_1, W + \xi_2)\, J_\vartheta(\xi_2, W)$$

for all $\xi_1, \xi_2 \in L$ and $W \in V$. We observe that for all $\xi_1, \xi_2 \in L$ and $W \in V$,

$$Q_{\xi_1+\xi_2}(W)+c_{\xi_1+\xi_2}\equiv Q_{\xi_1}(W+\xi_2)+Q_{\xi_2}(W)+c_{\xi_1}+c_{\xi_2} \mod \mathbb{Z}.$$

J_ϑ is called the automorphic factor of the theta function ϑ on V relative to L.

Theorem 2. *(Igusa [14], p.67). Let $J : L\times V \longrightarrow \mathbb{C}^\times$ be the automorphic factor of a theta function ϑ on V relative to L. Then there exists a unique triple (Q,ℓ,ψ) such that*

$$J(\xi,W)=e^{\pi\{Q(W,\xi)+\frac{1}{2}Q(\xi,\xi)+2i\,\ell(\xi)\}}\psi(\xi),\quad \xi\in L,\ W\in V, \tag{1.2.4}$$

where

1. *Q is a quasi-hermitian form on $V\times V$,*
2. *the hermitian form $H:=\mathrm{Her}\,(Q)$ defined by*

$$H(W_1,W_2)=\frac{1}{2i}\left\{Q(iW_1,W_2)-Q(W_1,iW_2)\right\},\quad W_1,\ W_2\in V$$

   is a Riemann form with respect to L, that is, $H={}^t\overline{H}>0$ and $(\mathrm{Im}\,H)(L\times L)\subset\mathbb{Z}$,
3. *$\ell : V\longrightarrow\mathbb{C}$ is a $\mathbb{C}$-linear form on V,*
4. *ψ is a second degree character of L which is associated with $A:=\mathrm{Im}\,H$,*
5. *ψ is strongly associated with A.*

Remark 1. (4) means that $\psi : L\longrightarrow\mathbb{C}_1^*$ is a semi-character of L satisfying the functional equation

$$\psi(\xi_1+\xi_2)=e^{\pi iA(\xi_1,\xi_2)}\psi(\xi_1)\psi(\xi_2),\quad \xi_1,\xi_2\in L. \tag{$*$}$$

Definition 3. A theta function with the automorphic factor of the form (1.2.4) is called a theta function of type (Q,ℓ,ψ). We denote by $L(Q,\ell,\psi)$ the union of theta functions of type (Q,ℓ,ψ) and the constant 0. A theta function of type (Q,ℓ,ψ) is said to be normalized if $\mathrm{Sym}\,Q=0$ and $\ell=0$. Here $\mathrm{Sym}\,Q : V\times V\longrightarrow\mathbb{C}$ is a symmetric $\mathbb{C}$-linear form on $V\times V$ defined by

$$(\mathrm{Sym}\,Q)(z,w)=\frac{1}{2i}\left\{Q(iz,w)+Q(z,iw)\right\},\quad z,w\in V.$$

We observe that $Q=\mathrm{Her}\,Q+\mathrm{Sym}\,Q$. We note that $\mathrm{Sym}\,Q=0$ if and only if $Q=\mathrm{Her}\,Q=H$. We denote by $Th(H,\psi,L)$ the union of the set of all normalized theta functions of type $(H,0,\psi)$ and the constant 0. It is easily seen that if $\vartheta\in Th(H,\psi,L)$, for all $W\in V,\ \xi\in L$, we have

$$\vartheta(W+\xi)=e^{\pi H(W+\frac{1}{2}\xi,\xi)}\psi(\xi)\,\vartheta(W). \tag{1.2.5}$$

Theorem 3. *Let S be a positive definite, symmetric integral matrix of degree m and let A,B be two $m\times n$ real matrices. Then for $\Omega\in\mathbb{H}_n$ and $W\in\mathbb{C}^{(m,n)}$, we have*

$$(\theta.1)\qquad \vartheta^{(S)}\begin{bmatrix}A\\B\end{bmatrix}(\Omega,-W)=\vartheta^{(S)}\begin{bmatrix}-A\\-B\end{bmatrix}(\Omega,W),$$

$$(\theta.2)\qquad \vartheta^{(S)}\begin{bmatrix}A\\B\end{bmatrix}(\Omega,W+\lambda\Omega+\mu)$$

$$=e^{-\pi i\,\sigma\{S(\lambda\Omega\,{}^t\lambda+2\,(W+\mu)\,{}^t\lambda)\}}\,e^{-2\pi i\,\sigma(SB\,{}^t\lambda)}\cdot\vartheta^{(S)}\begin{bmatrix}A+\lambda\\B+\mu\end{bmatrix}(\Omega,W)$$

for all $\lambda,\mu\in\mathbb{R}^{(m,n)}$.

$$(\theta.3)\qquad \vartheta^{(S)}\begin{bmatrix}A\\B\end{bmatrix}(\Omega,W)=e^{\pi i\,\sigma\{S(A\Omega\,{}^tA)+2\,(W+B)\,{}^tA)\}}\,\vartheta^{(S)}\begin{bmatrix}0\\0\end{bmatrix}(\Omega,W+A\Omega+B).$$

Moreover, if S *is a positive definite, symmetric integral matrix of degree* m, *we have*

$$(\theta.4)\qquad \vartheta^{(S)}\begin{bmatrix}A+\xi\\B+\eta\end{bmatrix}(\Omega,W)=e^{2\pi i\,\sigma(SA{}^t\eta)}\,\vartheta^{(S)}\begin{bmatrix}A\\B\end{bmatrix}(\Omega,W).$$

for all $\xi,\eta\in\mathbb{Z}^{(m,n)}$.

$$(\theta.5)\qquad \vartheta^{(S)}\begin{bmatrix}A\\B\end{bmatrix}(\Omega,W+\xi\Omega+\eta)$$

$$=e^{-\pi i\,\sigma\{S(\xi\Omega\,{}^t\xi+2\,W{}^t\xi)\}}\cdot e^{2\pi i\,\sigma\{S(A\,{}^t\eta-B\,{}^t\xi)\}}\cdot\vartheta^{(S)}\begin{bmatrix}A\\B\end{bmatrix}(\Omega,W)$$

for all $\xi,\eta\in\mathbb{Z}^{(m,n)}$.

Proof. $(\theta.1)$ follows immediately from the definition (1.2.3). $(\theta.2)$ follows immediately from the relation

$$\begin{aligned}&(N+A)\Omega\,{}^t(N+A)+2(W+\lambda\Omega+\mu+B)\,{}^t(N+A)\\=\;&(N+A+\lambda)\,\Omega\,{}^t(N+A+\lambda)+2(W+\mu+B)\,{}^t(N+A+\lambda)\\&-(N+A)\Omega\,{}^t\lambda+\lambda\Omega\,{}^t(N+A)-\lambda\Omega\,{}^t\lambda-2(W+\mu+B)\,{}^t\lambda.\end{aligned}$$

If we put $A=B=0$ and replace λ,μ by A,B in $(\theta.2)$, then we obtain $(\theta.3)$. For $\xi,\eta\in\mathbb{Z}^{(m,n)}$, we have

$$\begin{aligned}&\vartheta^{(S)}\begin{bmatrix}A+\xi\\B+\eta\end{bmatrix}(\Omega,W)\\=&\sum_{N\in\mathbb{Z}^{(m,n)}}e^{\pi i\,\sigma\{S((A+N+\xi)\Omega\,{}^t(A+N+\xi)+2\,(W+B)\,{}^t(A+N+\xi))\}}\\&\times e^{2\pi i\,\sigma\{S\eta\,{}^t(N+\xi)\}}\cdot e^{2\pi i\,\sigma(S\,{}^t\eta A)}\\=\;&e^{2\pi i\,\sigma(SA\,{}^t\eta)}\cdot\vartheta^{(S)}\begin{bmatrix}A\\B\end{bmatrix}(\Omega,W).\end{aligned}$$

Here in the last equality we used the fact that $\sigma(S\eta\,{}^t(N+\xi)) \in \mathbb{Z}$ because S is integral. (θ.5) follows from (θ.2), (θ.4) and the fact that $\sigma(S\eta\,{}^t\xi)$ is integral. □

For a positve definite, symmetric real matrix S of degree m, $\Omega \in \mathbb{H}_n$ and $A, B \in \mathbb{R}^{(m,n)}$, we put

$$\chi_{S,\Omega,A,B}(\xi\Omega + \eta) := \chi_{S,\Omega,A,B}(\xi,\eta) := e^{2\pi i\,\sigma\{S(A\,{}^t\eta - B\,{}^t\xi)\}}, \tag{1.2.6}$$

where $\xi, \eta \in \mathbb{Z}^{(m,n)}$.

We define

$$q_{S,\Omega}(W) = \frac{1}{2}\,\sigma\left(SW(\Omega - \overline{\Omega})^{-1}\,{}^tW\right), \quad W \in \mathbb{C}^{(m,n)} \tag{1.2.7}$$

and also define

$$H_{S,\Omega}(W_1, W_2) = 2\,i\,\sigma\left(SW_1(\Omega - \overline{\Omega})^{-1}\,{}^t\overline{W}_2\right), \quad W_1, W_2 \in \mathbb{C}^{(m,n)}. \tag{1.2.8}$$

It is easy to check that $H_{S,\Omega}$ is a positive hermitian form on $\mathbb{C}^{(m,n)}$.

Lemma 3. *For $W \in \mathbb{C}^{(m,n)}$ and $l \in \mathbb{Z}^{(m,n)}\Omega + \mathbb{Z}^{(m,n)}$, we have*

$$q_{S,\Omega}(W + l) = q_{S,\Omega}(W) + q_{S,\Omega}(l) + \sigma\left(Sl(\Omega - \overline{\Omega})^{-1}\,{}^tW\right) \tag{1.2.9}$$

and

$$\begin{aligned} H_{S,\Omega}\left(W + \frac{l}{2}, l\right) &= \sigma\left(S\left(W + \tfrac{l}{2}\right)(\operatorname{Im}\Omega)^{-1}\,{}^tl\right) \\ &\quad -2\,i\,\sigma\left(S\left(W + \tfrac{l}{2}\right)\,{}^t\xi\right), \end{aligned} \tag{1.2.10}$$

where $l = \xi\Omega + \eta$, $\xi, \eta \in \mathbb{Z}^{(m,n)}$.

Proof. It follows immediately from a straightforward computation. □

Lemma 4. *Let S be a positive definite, symmetric integral matrix of degree m. For $\Omega \in \mathbb{H}_n$, we let $L_\Omega := \mathbb{Z}^{(m,n)}\Omega + \mathbb{Z}^{(m,n)}$ be the lattice in $\mathbb{C}^{(m,n)}$. We define the mapping $\psi_{S,\Omega} : L_\Omega \longrightarrow \mathbb{C}_1^*$ by*

$$\psi_{S,\Omega}(\xi\Omega + \eta) = e^{\pi i\,\sigma(S\eta\,{}^t\xi)}, \quad \xi, \eta \in \mathbb{Z}^{(m,n)}. \tag{1.2.11}$$

Then

(a) *$\psi_{S,\Omega}$ is a second-degree character of L_Ω associated with* $\operatorname{Im} H_{S,\Omega}$.

(b) *$\psi_{S,\Omega} \cdot \chi_{S,\Omega,A,B}$ is a second-degree character of L_Ω associated with* $\operatorname{Im} H_{S,\Omega}$.

Proof. (a) We fix $l = \xi\Omega + \eta \in L_\Omega$ with $\xi, \eta \in \mathbb{Z}^{(m,n)}$. We define $f_l : L_\Omega \longrightarrow \mathbb{C}_1^\times$ by

$$f_l(l_1) := \frac{\psi_{S,\Omega}(l_1 + l)}{\psi_{S,\Omega}(l_1)\,\psi_{S,\Omega}(l)}\ , \quad l_1 \in L_\Omega.$$

It is easy to see that f_l is a character of L_Ω and hence to see that the map from $L_\Omega \times L_\Omega$ to $\mathbb{C}_1^*$ defined by

$$(l_1, l_2) \longmapsto \frac{\psi_{S,\Omega}(l_1 + l_2)}{\psi_{S,\Omega}(l_1)\,\psi_{S,\Omega}(l_2)}$$

is a bicharacter of L_Ω, i.e., a character of L_Ω in l_1 and l_2. Hence $\psi_{S,\Omega}$ is a second degree character of L_Ω. In order to show that $\psi_{S,\Omega}$ is associated with $H_{S,\Omega}$, it is enough to prove that

$$\psi_{S,\Omega}(l_1 + l_2) = e^{\pi i A_{S,\Omega}(l_1,l_2)}\,\psi_{S,\Omega}(l_1)\,\psi_{S,\Omega}(l_2) \tag{1.2.12}$$

for all $l_1, l_2 \in L_\Omega$. Here $A_{S,\Omega}$ denotes the imaginary part of the positive hermitian form $H_{S,\Omega}$. By an easy computation, we have

$$A_{S,\Omega}(l_1, l_2) = \sigma\{S(\xi_1\,{}^t\eta_2 - \eta_1\,{}^t\xi_2)\}, \tag{1.2.13}$$

where $l_i = \xi_i\Omega + \eta_i \in L_\Omega\,(1 \le i \le 2)$. Hence (1.2.12) follows immediately from (1.2.13).

(b) We fix $l = \xi\Omega + \eta \in L_\Omega$ with $\xi, \eta \in \mathbb{Z}^{(m,n)}$. We put $\widetilde{\psi}_{S,\Omega,A,B} := \psi_{S,\Omega} \cdot \chi_{S,\Omega,A,B}$. Then the map $\tilde{f}_l : L_\Omega \longrightarrow \mathbb{C}_1^*$ defined by

$$\tilde{f}_l(l_1) = \frac{\widetilde{\psi}_{S,\Omega,A,B}(l_1 + l)}{\widetilde{\psi}_{S,\Omega,A,B}(l_1)\,\widetilde{\psi}_{S,\Omega,A,B}(l_2)}\ , \quad l_1 \in L_\Omega$$

is a character of L_Ω. So $\widetilde{\psi}_{S,\Omega,A,B}$ is a second degree character of L_Ω. In order to show that $\widetilde{\psi}_{S,\Omega}$ is associated with $A_{S,\Omega}$, it suffices to prove that

$$\widetilde{\psi}_{S,\Omega,A,B}(l_1 + l_2) = e^{\pi i A_{S,\Omega}(l_1,l_2)}\,\widetilde{\psi}_{S,\Omega,A,B}(l_1)\,\widetilde{\psi}_{S,\Omega,A,B}(l_2) \tag{1.2.14}$$

for all $l_1, l_2 \in L_\Omega$. An easy calculation yields (1.2.14). □

Theorem 4. *We assume that S is a positive definite, symmetric integral matrix of degree m. Let $\Omega \in \mathbb{H}_n$. We denote by R_S^Ω the vector space of all holomorphic functions $f : \mathbb{C}^{(m,n)} \longrightarrow \mathbb{C}$ satisfying the transformation behaviour*

$$f(W + \xi\Omega + \eta) = e^{-\pi i\,\sigma\{S(\xi\Omega\,{}^t\xi + 2\,\xi\,{}^tW)\}}\,f(W), \quad W \in \mathbb{C}^{(m,n)}$$

for all $\xi, \eta \in \mathbb{Z}^{(m,n)}$. Then the mapping

$$\Theta : R_S^\Omega \longrightarrow Th(H_{S,\Omega}, \psi_{S,\Omega}, L_\Omega)$$

defined by

$$(\Theta(F))(W) := e^{2\pi i\, q_{S,\Omega}(W)} F(W), \quad F \in R_S^\Omega,\ W \in \mathbb{C}^{(m,n)}$$

is an isomorphism of vector spaces, where L_Ω and $\psi_{S,\Omega}$ are the same as in Lemma 4.

Proof. First we will show the image $\Theta(R_S^\Omega)$ is contained in $Th(H_{S,\Omega}, \psi_{S,\Omega}, L_\Omega)$. If $F \in R_S^\Omega,\ W \in \mathbb{C}^{(m,n)}$ and $l = \xi\Omega + \eta \in L_\Omega$, then we have

$$\begin{aligned}
\Theta(F)(W+l) &= e^{2\pi i\, q_{S,\Omega}(W+l)}\, F(W+l) \\
&= e^{2\pi i \{q_{S,\Omega}(W) + q_{S,\Omega}(l) + \sigma(Sl(\Omega - \overline{\Omega}^{-1}\, {}^tW)\}} \\
&\quad \times e^{-\pi i\, \sigma\{S(\xi\Omega\, {}^t\xi + 2\, W\, {}^t\xi)\}}\, F(W) \qquad (\text{by Lemma 3}) \\
&= e^{2\pi i\, \sigma\{S(W+\frac{l}{2})(\Omega - \overline{\Omega})^{-1}\, {}^tl\}} \\
&\quad \times e^{-\pi i\, \sigma\{S(\xi\Omega\, {}^t\xi + 2\, W\, {}^t\xi + 2\, W\, {}^t\xi)\}} \cdot \Theta(F)(W) \\
&= e^{\pi H_{S,\Omega}(W+\frac{l}{2}, l)} \cdot e^{-\pi i\, \sigma(S\eta\, {}^t\xi)}\, \Theta(F)(W) \\
&= e^{\pi H_{S,\Omega}(W+\frac{l}{2}, l)}\, \psi_{S,\Omega}(l)\, \Theta(F)(W).
\end{aligned}$$

Thus $\Theta(F)$ is contained in the set $Th(H_{S,\Omega}, \psi_{S,\Omega}, L_\Omega)$. It is easy to see that the mapping Θ is an isomorphism. □

Proposition 1. *Let S be as above in Theorem 4 and $A, B \in \mathbb{R}^{(m,n)}$. We denote by $R_{S,A,B}^\Omega$ the union of the set of all theta functions with characteristic (A, B) with respect to S and Ω and the constant 0. Then we have an isomorphism*

$$R_{S,A,B}^\Omega \cong Th(H_{S,\Omega}, \psi_{S,\Omega} \cdot \chi_{S,\Omega,A,B}, L_\Omega).$$

Proof. First we observe that $\psi_{S,\Omega} \cdot \chi_{S,\Omega,A,B}$ is a second degree character of L_Ω associated with $A_{S,\Omega}$ (cf. Lemma 4 (B)). In a similar way in the proof of Theorem 4, using $(\theta.5)$, we can show that the mapping

$$\Theta_{A,B}(f)(W) := e^{2\pi i\, q_{S,\Omega}(W)}\, f(W), \quad f \in R_{S,A,B}^\Omega,\ W \in \mathbb{C}^{(m,n)}$$

has its image in $Th(H_{S,\Omega}, \psi_{S,\Omega} \cdot \chi_{S,\Omega,A,B}, L_\Omega)$. □

Proposition 2. *Let S be as above in Theorem 4 and let $A, B \in \mathbb{R}^{(m,n)}$. Then we have an isomorphism*

$$Th(H_{S,\Omega}, \psi_{S,\Omega}, L_\Omega) \cong TH(H_{S,\Omega}, \psi_{S,\Omega} \cdot \chi_{S,\Omega,A,B}, L_\Omega).$$

Proof. The proof follows from the fact that the dimension of the complex vector space $Th(H_{S,\Omega}, \psi_{S,\Omega}, L_\Omega)$ is equal to that of $Th(H_{S,\Omega}, \psi_{S,\Omega} \cdot \chi_{S,\Omega,A,B}, L_\Omega)$. It is well known that the dimension of $Th(H_{S,\Omega}, \psi_{S,\Omega}, L_\Omega)$ is equal to the Pfaffian of $A_{S,\Omega}$ relative to L_Ω (cf. [14], p.72). □

Remark 2. From Theorem 4, Proposition 1 and Proposition 2, R_S^Ω is isomorphic to $R_{S,A,B}^\Omega$ for any $A, B \in \mathbb{R}^{(m,n)}$.

Now as before, we fix an element $\Omega \in \mathbb{H}_n$ and let S be a positive symmetric integral matrix of degree m. Then the lattice $L := \mathbb{Z}^{(m,n)} \times \mathbb{Z}^{(m,n)}$ acts on $\mathbb{C}^{(m,n)}$ freely by

$$(\xi,\,\eta)\cdot W = W + \xi\Omega + \eta, \quad \xi,\eta \in \mathbb{Z}^{(m,n)}, \quad W \in \mathbb{C}^{(m,n)}.$$

Lemma 5. *Let* $A,\, B \in \mathbb{R}^{(m,n)}$. *Let* $J_{S,\Omega,A,B} : L \times \mathbb{C}^{(m,n)} \longrightarrow \mathbb{C}^*$ *be the mapping defined by*

$$J_{S,\Omega,A,B}(l,W) := e^{\pi i\,\sigma\{S(\xi\Omega\,{}^t\xi + 2\,W\,{}^t\xi)\}}\cdot e^{-2\pi i\,\sigma\{S(A\,{}^t\eta - B\,{}^t\xi)\}}, \tag{1.2.15}$$

where $l = (\xi,\eta) \in L$ *and* $W \in \mathbb{C}^{(m,n)}$. *Then* $J_{S,\Omega,A,B}$ *is an automorphic factor for the lattice* L.

Proof. For brevity, we write $J := J_{S,\Omega,A,B}$. For any two elements $l_i = (\xi_i,\,\eta_i)\,(\,i=1,2\,)$ of L and $W \in \mathbb{C}^{(m,n)}$, we must show that

$$J(l_1 + l_2,\, W) = J(l_1,\, l_2 + W)\, J(l_2, W). \tag{1.2.16}$$

Using the fact that $\sigma(2S\eta_2\,{}^t\xi_1)$ is an even integer, an easy computation yields (1.2.16). □

The Heisenberg group $H_{\mathbb{R}}^{(n,m)}$ with multiplication $\diamond$ acts on $\mathbb{C}^{(m,n)}$ by

$$[\lambda_0,\mu_0,\kappa_0]\cdot(\lambda\Omega+\mu) := (\lambda_0+\lambda)\Omega + (\mu_0+\mu), \quad \lambda,\mu \in \mathbb{R}^{(m,n)}.$$

Since the center $\mathcal{Z} = \left\{\,[0,0,\kappa] \mid \kappa = {}^t\kappa \in \mathbb{R}^{(m,m)}\,\right\}$ of $H_{\mathbb{R}}^{(n,m)}$ is the stabilizer of $H_{\mathbb{R}}^{(n,m)}$ at 0, the homogeneous space $H_{\mathbb{R}}^{(n,m)}/\mathcal{Z}$ is identified with $\mathbb{C}^{(m,n)}$ via

$$[\lambda,\mu,\kappa]\cdot\mathcal{Z} \longmapsto [\lambda,\mu,\kappa]\cdot 0 = \lambda\Omega + \mu.$$

Thus the automorphic factor $J_{S,\Omega,A,B}$ for the lattice L may be lifted to the automorphic factor $\tilde{J}_{S,\Omega,A,B} : H_{\mathbb{R}}^{(n,m)} \times \mathbb{C}^{(m,n)} \longrightarrow \mathbb{C}^*$ defined by

$$\tilde{J}_{S,\Omega,A,B}(g_0,W) = e^{\pi i\,\sigma\{S(\lambda\Omega\,{}^t\lambda + 2\,W\,{}^t\lambda + \kappa)\}}\cdot e^{-\pi i\,\sigma\{S(A\,{}^t\mu - B\,{}^t\lambda)\}}, \tag{1.2.17}$$

where $g_0 = [\lambda,\mu,\kappa] \in H_{\mathbb{R}}^{(n,m)}$.

We denote by $\mathcal{A}_{S,\Omega}$ be the complex vector space consisting of $\mathbb{C}$-valued smooth functions φ on $H_{\mathbb{R}}^{(n,m)}$ satisfying the following conditions

(a) $\varphi([\xi,\eta,0]\diamond g_0) = \varphi(g_0)$ for all $\xi,\eta \in \mathbb{Z}^{(m,n)}$ and $g_0 \in H_{\mathbb{R}}^{(n,m)}$,

(b) $\varphi(g_0 \diamond [0,0,\kappa]) = e^{\pi i\,\sigma(S\kappa)}\,\varphi(g_0)$ for all $\kappa = {}^t\kappa \in \mathbb{R}^{(m,m)}$ and $g_0 \in H_{\mathbb{R}}^{(n,m)}$,

(c) $\left(\mathcal{L}_{X_{ka}} - \sum_{b=1}^{n}\Omega\mathcal{L}_{\widehat{X}_{kb}}\right) = 0$ for all $1 \le k \le m$ and $1 \le a \le n$.

Here if X is an element of the Lie algebra of $H_{\mathbb{R}}^{(n,m)}$,

$$\big(\mathcal{L}_X\varphi\big)(g_0) = \frac{d}{dt}\bigg|_{t=0}\varphi(g_0 \diamond \exp tX), \quad g_0 \in H_{\mathbb{R}}^{(n,m)}.$$

Theorem 5. *Let S and Ω be as before. Then the vector space R_S^Ω is isomorphic to the vector space $\mathcal{A}_{S,\Omega}$ via the mapping*

$$f \longmapsto \varphi_f(g_0) := \tilde{J}_{S,\Omega,0,0}(g_0,\, 0)\, f(g_0 \cdot 0),$$

where $g_0 \in H_{\mathbb{R}}^{(n,m)}$ and $f \in R_S^\Omega$.

The inverse of the above isomorphism is given by

$$\varphi \longmapsto f_\varphi(W) := \tilde{J}_{S,\Omega,0,0}(g_0,\, 0)^{-1}\,\varphi(g_0), \quad \varphi \in \mathcal{A}_{S,\Omega},$$

where $W = g_0 \cdot 0$. This definition does not depend on the choice of g_0 with $W = g_0 \cdot 0$.

Proof. For brevity, we write $\tilde{J} := \tilde{J}_{S,\Omega,0,0}$. If $\gamma = [\xi, \eta, 0] \in H_{\mathbb{R}}^{(n,m)}$ with $\xi, \eta \in \mathbb{Z}^{(m,n)}$, we have for all $g_0 \in H_{\mathbb{R}}^{(n,m)}$

$$\begin{aligned}
\varphi_f(\gamma \diamond g_0) =& \tilde{J}(\gamma \diamond g_0,\, 0)\, f((\gamma \diamond g_0) \cdot 0)\\
=& \tilde{J}(\gamma,\, g_0 \cdot 0)\, \tilde{J}(g_0,\, 0)\, f(g_0 \cdot 0 + \xi\Omega + \eta)\\
=& \tilde{J}(\gamma,\, g_0 \cdot 0)\, \tilde{J}(g_0, 0)\, J((\xi,\eta),\, g_0 \cdot 0)^{-1}\, f(g_0 \cdot 0)\\
=& \tilde{J}(\gamma, 0)\, f(g_0 \cdot 0)\\
=& \varphi_f(g_0).
\end{aligned}$$

And if $\kappa = {}^t\kappa \in \mathbb{R}^{(m,m)}$, we have

$$\begin{aligned}
\varphi_f(g_0 \diamond [0,0,\kappa]) &= \tilde{J}(g_0 \diamond [0,0,\kappa],\, 0)\, f((g_0 \diamond [0,0,\kappa]) \cdot 0)\\
&= \tilde{J}(g_0,\, [0,0,\kappa] \cdot 0)\, \tilde{J}([0,0,\kappa],\, 0)\, f(g_0 \cdot 0)\\
&= e^{\pi i\, \sigma(S\kappa)}\, \tilde{J}(g_0, 0)\, f(g_0 \cdot 0)\\
&= e^{\pi \sigma(S\kappa)}\, \varphi_f(g_0).
\end{aligned}$$

We introduce a system of complex coordinates on $\mathbb{C}^{(m,n)}$ with respect to Ω:

$$W = \lambda\Omega + \mu, \quad \overline{W} = \lambda\overline{\Omega} + \mu, \quad \lambda, \mu \text{ real}.$$

We set

$$dW = \begin{pmatrix} dW_{11} & dW_{12} & \dots & dW_{1n}\\ dW_{21} & dW_{22} & \dots & dW_{2n}\\ \vdots & \vdots & \ddots & \vdots\\ dW_{m1} & dW_{m2} & \dots & dW_{mn} \end{pmatrix}, \quad \frac{\partial}{\partial W} = \begin{pmatrix} \frac{\partial}{\partial W_{11}} & \frac{\partial}{\partial W_{21}} & \cdots & \frac{\partial}{\partial W_{m1}}\\ \frac{\partial}{\partial W_{12}} & \frac{\partial}{\partial W_{22}} & \cdots & \frac{\partial}{\partial W_{m2}}\\ \vdots & \vdots & \ddots & \vdots\\ \frac{\partial}{\partial W_{1n}} & \frac{\partial}{\partial W_{2n}} & \cdots & \frac{\partial}{\partial W_{mn}} \end{pmatrix}.$$

Then an easy computation yields

$$\begin{aligned}\frac{\partial}{\partial\lambda} &= \Omega\frac{\partial}{\partial W} + \overline{\Omega}\frac{\partial}{\partial\overline{W}},\\ \frac{\partial}{\partial\mu} &= \frac{\partial}{\partial W} + \frac{\partial}{\partial\overline{W}}.\end{aligned}$$

Thus we obtain the following

$$\frac{\partial}{\partial\overline{W}} = \frac{i}{2}\,(\operatorname{Im}\Omega)^{-1}\left(\frac{\partial}{\partial\lambda} - \Omega\frac{\partial}{\partial\mu}\right). \tag{1.2.18}$$

Since f is holomorphic, according to (1.2.18), f satisfies the conditions

$$\left(\frac{\partial}{\partial\lambda_{ka}} - \sum_{b=1}^{n}\Omega_{ab}\frac{\partial}{\partial\mu_{kb}}\right) f(W) = 0, \quad 1\le k\le m,\ 1\le a\le n. \tag{1.2.19}$$

Conversely, if a smooth function on $\mathbb{C}^{(m,n)}$ satisfies the condition (1.2.19), it is holomorphic.

In order to prove that φ_f satisfies the condition (c), we first compute $\mathcal{L}_{X_{ka}}\varphi_f$ and $\mathcal{L}_{\widehat{X}_{lb}}\varphi_f$ for $1\le k,l\le m$ and $1\le a,b\le n$. If $g=[\lambda,\mu,\kappa]\in H_{\mathbb{R}}^{(n,m)}$ and $S=(s_{kl})$,

$$\begin{aligned}\left(\mathcal{L}_{X_{ka}}\varphi_f\right) &= \frac{d}{dt}\bigg|_{t=0}\varphi_f(g\diamond\exp tX_{ka})\\ &= \frac{d}{dt}\bigg|_{t=0}\varphi_f([\lambda,\mu,\kappa]\diamond[tE_{ka},0,0])\\ &= \frac{d}{dt}\bigg|_{t=0}\tilde{J}([\lambda+tE_{ka},\mu,\kappa],\,0)\,f((\lambda+tE_{ka})\Omega+\mu)\\ &= \frac{d}{dt}\bigg|_{t=0} e^{\pi i\,\sigma\{S(\lambda+tE_{ka})\Omega\,{}^t(\lambda+tE_{ka})\}}\,e^{\pi i\,\sigma(S\kappa)}\,f((\lambda+tE_{ka})\Omega+\mu)\\ &= e^{\pi i\,\sigma\{S(\kappa+\lambda\Omega\,{}^t\lambda)\}}\left\{2\pi i\left(\sum_{b=1}^{n}\sum_{l=1}^{m}s_{kl}\Omega_{ab}\lambda_{ab}\right)+\frac{\partial}{\partial\lambda_{ka}}\right\}f(W).\end{aligned}$$

On the other hand,

$$\begin{aligned}
\left(\mathcal{L}_{\widehat{X}_{lb}}\varphi_f\right)(g) =& \frac{d}{dt}\bigg|_{t=0} \varphi_f(g \diamond \exp t\widehat{X}_{lb}) \\
=& \frac{d}{dt}\bigg|_{t=0} \varphi_f([\lambda,\mu,\kappa] \diamond [0, tE_{lb}, 0]) \\
=& \frac{d}{dt}\bigg|_{t=0} \varphi_f([\lambda, \mu + tE_{lb}, \kappa + t\lambda\, {}^tE_{kb} + tE_{kb}\, {}^t\lambda]) \\
=& e^{\pi i\, \sigma\{S(\kappa+\lambda\Omega\, {}^t\lambda)\}} \frac{d}{dt}\bigg|_{t=0} e^{2\pi i t\, \sigma(S\lambda\, {}^tE_{lb})} f(\lambda\Omega + (\mu + tE_{lb})) \\
=& e^{\pi i\, \sigma\{S(\kappa+\lambda\Omega\, {}^t\lambda)\}} \left\{ 2\pi i \left(\sum_{p=1}^{m} s_{lp}\lambda_{pb} \right) + \frac{\partial}{\partial \mu_{lb}} \right\} f(W).
\end{aligned}$$

Thus

$$\begin{aligned}
&\Big(\mathcal{L}_{X_{ka}} - \sum_{b=1}^{n} \Omega_{ab}\mathcal{L}_{\widehat{X}_{kb}}\Big)\varphi_f(g) \\
=& e^{\pi i\, \sigma\{S(\kappa+\lambda\Omega\, {}^t\lambda)\}} \left\{ \frac{\partial}{\partial \lambda_{ka}} - \sum_{b=1}^{n} \Omega_{ab}\frac{\partial}{\partial \mu_{kb}} \right\} f(W) = 0.
\end{aligned}$$

This completes the proof. □

1.3 Induced Representations

Let G be a locally compact, separable topological group and K be a closed subgroup of G. Let σ be an irreducible unitary representation of K in a separable Hilbert space $\mathcal{H}$. Let μ be a G-invariant measure in the homogeneous space $X := K\backslash G = \{Kg \mid g \in G\}$ of the right K-cosets in G. We denote the induced representation of G from σ by

$$U_\sigma := \operatorname{Ind}_K^G \sigma.$$

Let $\mathcal{H}^\sigma$ be the Hilbert space consisting of all functions $\phi : G \longrightarrow \mathcal{H}$ which satisfy the following conditions:

(1) $(\phi(g), v)_{\mathcal{H}}$ is measurable with respect to dg for all $v \in \mathcal{H}$.

(2) $\phi(kg) = \sigma(k)(\phi(g))$ for all $k \in K$ and $g \in G$.

(3) $\| \phi \|^2 = \int_X \| \phi(g) \|^2 \, d\mu(\dot{g}) < \infty, \quad \dot{g} = Kg,$

where dg is a G-invariant measure on G and $(\ , \ )_{\mathcal{H}}$ is an inner product in $\mathcal{H}$ and $\| \phi(g) \|$ is the norm in $\mathcal{H}$. The inner product $(\ , \ )$ in $\mathcal{H}^\sigma$ is given by

$$(\phi_1, \phi_2) = \int_X (\phi_1(g), \phi_2(g))_{\mathcal{H}} \, d\mu(\dot{g}), \quad \phi_1, \phi_2 \in \mathcal{H}^\sigma.$$

Then $U_\sigma = \operatorname{Ind}_K^G \sigma$ is realized in the Hilbert space $\mathcal{H}^\sigma$ as follows:

$$\left(U_\sigma(g_0)\phi\right)(g) = \phi(gg_0), \quad g, g_0 \in G, \ \phi \in \mathcal{H}^\sigma. \tag{1.3.1}$$

It is easy to see that $\mathcal{H}^\sigma$ is isomorphic to the Hilbert space $\mathcal{H}_\sigma := L^2(X, \mu, \mathcal{H})$ of square integrable functions $f : X \longrightarrow \mathcal{H}$ with values in $\mathcal{H}$ via the formula

$$\phi_f(g) = \sigma(k_g)\left(f(\dot{g})\right), \quad f \in \mathcal{H}_\sigma, \ g \in G, \tag{1.3.2}$$

where $\dot{g} = Kg$ and k_g is the K-component of g in the Mackey decomposition $g = k_g s_g$.

We can show easily that U_σ is realized in $\mathcal{H}_\sigma$ by

$$\left(U_\sigma(g_0)f\right)(\dot{g}) = \sigma(k_{s_g g_0})\left(f(\dot{g}g_0)\right), \quad g_0 \in G, \ f \in \mathcal{H}_\sigma, \ \dot{g} = Kg \in X, \tag{1.3.3}$$

where $k_{s_g g_0}$ denotes the K-component of $s_g g_0$ in the Mackey decomposition of $s_g g_0$.

If σ is a one-dimensional representation of K, U_σ is called a monomial representation.

Remark 3. It is interesting to find out irreducible closed subspaces of $\mathcal{H}^\sigma$ or $\mathcal{H}_\sigma$ invariant under G.

We recall $\mathcal{A}$, $\mathcal{S}$, G_H, $\mathcal{S}_{\hat{\kappa}}$ etc in Section 1.1. Mackey's method teaches us that an irreducible unitary representation of $G_H \cong H_{\mathbb{R}}^{(n,m)}$ is of the following form

$$T_{\hat{\kappa}} := \mathrm{Ind}_{\mathcal{S}_{\hat{\kappa}} \ltimes \mathcal{A}}^{G_H} \chi_{\hat{\kappa}} \cdot \hat{a}(\hat{\kappa}) = \mathrm{Ind}_{\mathcal{A}}^{G_H} \hat{a}(\hat{\kappa})$$

or

$$T_{\hat{x},\hat{y}} := \mathrm{Ind}_{\mathcal{S}_{\hat{y}} \ltimes \mathcal{A}}^{G_H} \chi_{\hat{x}} \cdot \hat{a}(\hat{y}) = \mathrm{Ind}_{\mathcal{S} \ltimes \mathcal{A}}^{G_H} \chi_{\hat{x}} \cdot \hat{a}(\hat{y}),$$

where $\chi_{\hat{x}}$ is the character of $\mathcal{S}$ defined by $\chi_{\hat{x}}(l) := e^{2\pi i \sigma(\hat{x}\, {}^t l)}$ for $l \in \mathcal{S}$. Therefore the unitary dual $\widehat{G}_H$ of G_H or $H_{\mathbb{R}}^{(n,m)}$ is determined completely by

Type I. $\hat{\kappa} \in \mathrm{Sym}\,(m, \mathbb{R}),\quad \hat{\kappa} \neq 0.$
Type II. $(\hat{x}, \hat{y}) \in \mathbb{R}^{(m,n)} \times \mathbb{R}^{(m,n)}$ with $\hat{x}, \hat{y} \in \mathbb{R}^{(m,n)}$.

The representation $\rho \in \widehat{G}_H$ of *type* **I** acts nontrivially on the center $\mathcal{Z} \cong \mathrm{Sym}\,(m, \mathbb{R})$ of G_H. On the other hand, the representation $\rho \in \widehat{G}_H$ of *type* **II** acts trivially on the center $\mathcal{Z}$ of G_H.

1.4 The Schrödinger Representation

For two fixed positive integers m and n, we put $G := H_{\mathbb{R}}^{(n,m)}$ and

$$K := \left\{ (0, \mu, \kappa) \in G \mid \mu \in \mathbb{R}^{(m,n)},\ \kappa = {}^t\kappa \in \mathbb{R}^{(m,m)} \right\}. \tag{1.4.1}$$

We note that $K = A$ (cf. Section 1.1, (1.1.4)) and that K is a closed, commutative normal subgroup of G. Since $(\lambda, \mu, \kappa) = (0, \mu, \kappa + \mu\, {}^t\lambda) \circ (\lambda, 0, 0)$ for $(\lambda, \mu, \kappa) \in G$, the homogeneous space $X := K\backslash G$ is identified with $\mathbb{R}^{(m,n)}$ via

$$Kg = K \circ (\lambda, 0, 0) \longmapsto \lambda, \quad g = (\lambda, \mu, \kappa) \in G.$$

We observe that G acts on X by

$$(Kg) \cdot g_0 = K\,(\lambda + \lambda_0, 0, 0), \tag{1.4.2}$$

where $g = (\lambda, \mu, \kappa) \in G$ and $g_0 = (\lambda_0, \mu_0, \kappa_0) \in G$.

If $g = (\lambda, \mu, \kappa) \in G$, we have

$$k_g = (0, \mu, \kappa + \mu\, {}^t\lambda), \quad s_g = (\lambda, 0, 0) \tag{1.4.3}$$

in the Mackey decomposition of $g = k_g \circ s_g$. Thus if $g_0 = (\lambda_0, \mu_0, \kappa_0) \in G$, then we have

$$s_g \circ g_0 = (\lambda, 0, 0) \circ (\lambda_0, \mu_0, \kappa_0) = (\lambda + \lambda_0, \mu_0, \kappa_0 + \lambda\, {}^t\mu_0) \tag{1.4.4}$$

and so

$$k_{s_g \circ g_0} = (0, \mu_0, \kappa_0 + \mu_0\, {}^t\lambda_0 + \lambda\, {}^t\mu_0 + \mu_0\, {}^t\lambda). \tag{1.4.5}$$

For a real symmetric matrix $c = {}^tc \in \mathbb{R}^{(m,m)}$ with $c \neq 0$, we consider the one-dimensional unitary representation σ_c of K defined by

$$\sigma_c\,((0, \mu, \kappa)) := e^{2\pi i\, \sigma(c\kappa)}\, I, \quad (0, \mu, \kappa) \in K, \tag{1.4.6}$$

where I denotes the identity mapping. Then the induced representation $U_{\sigma_c} := \mathrm{Ind}_K^G\, \sigma_c$ of G induced from σ_c is realized in the Hilbert space $\mathcal{H}_{\sigma_c} = L^2(X, d\dot{g}, \mathbb{C}) \cong L^2\left(\mathbb{R}^{(m,n)}, d\xi\right)$ as follows. If $g_0 = (\lambda_0, \mu_0, \kappa_0) \in G$ and $x = Kg \in X$ with $g = (\lambda, \mu, \kappa) \in G$, then according to (1.3.3), we have

$$\left(U_{\sigma_c}(g_0) f\right)(x) = \sigma_c\left(k_{s_g \circ g_0}\right)\left(f(x g_0)\right), \quad f \in \mathcal{H}_{\sigma_c}. \tag{1.4.7}$$

It follows from (1.4.5) that

$$\left(U_{\sigma_c}(g_0) f\right)(\lambda) = e^{2\pi i \sigma\{c(\kappa_0 + \mu_0\, {}^t\lambda_0 + 2\lambda\, {}^t\mu_0)\}}\, f(\lambda + \lambda_0). \tag{1.4.8}$$

Here we identified $x = Kg$ (resp. $xg_0 = Kgg_0$) with λ (resp. $\lambda + \lambda_0$). The induced representation U_{σ_c} is called the Schrödinger representation of G associated with σ_c. U_{σ_c} is a monomial representation.

In the previous section, we denoted by $\mathcal{H}^{\sigma_c}$ the Hilbert space consisting of all functions $\phi : G \longrightarrow \mathbb{C}$ which satisfy the following conditions:

(1) $\phi(g)$ is measurable with respect to dg.

(2) $\phi\left((0,\mu,\kappa)\circ g)\right) = e^{2\pi i\,\sigma(c\kappa)}\phi(g)$ for all $g \in G$.

(3) $\| \phi \|^2 := \int_X |\phi(g)|^2\, d\dot{g} < \infty, \quad \dot{g} = Kg,$

where dg (resp. $d\dot{g}$) is a G-invariant measure on G (resp. $X = K\backslash G$). The inner product $(\ ,\ )$ on $\mathcal{H}^{\sigma_c}$ is given by

$$(\phi_1, \phi_2) = \int_G \phi_1(g)\,\overline{\phi_2(g)}\, dg, \quad \phi_1,\ \phi_2 \in \mathcal{H}^{\sigma_c}.$$

We observe that $\mathcal{H}_{\sigma_c} = L^2(\mathbb{R}^{(m,n)}, d\xi)$ and that the mapping $\Phi_c : \mathcal{H}_{\sigma_c} \longrightarrow \mathcal{H}^{\sigma_c}$ defined by

$$\left(\Phi_c(f)\right)(g) := \phi_f(g) := e^{2\pi i\sigma\{c(\kappa+\mu\,{}^t\lambda)\}}\, f(\lambda) \tag{1.4.9}$$

$(f \in \mathcal{H}_{\sigma_c},\ g = (\lambda,\mu,\kappa) \in G)$ is an isomorphism of Hilbert spaces. The inverse $\Psi_c : \mathcal{H}^{\sigma_c} \longrightarrow \mathcal{H}_{\sigma_c}$ of Φ_c is given by

$$\left(\Psi_c(\phi)\right)(\lambda) := f_\phi(\lambda) := \phi((\lambda,0,0)), \quad \phi \in \mathcal{H}^{\sigma_c},\ \lambda \in \mathbb{R}^{(m,n)}. \tag{1.4.10}$$

From now on, for brevity we put

$$U_c = U_{\sigma_c}, \quad \mathcal{H}_c = \mathcal{H}_{\sigma_c} \quad \text{and} \quad \mathcal{H}^c = \mathcal{H}^{\sigma_c}.$$

The Schrödinger representation U_c of G on $\mathcal{H}^c$ is given by

$$\left(U_c(g_0)\phi\right)(g) = e^{2\pi i\sigma\{c(\kappa_0+\mu_0\,{}^t\lambda_0+\lambda\,{}^t\mu_0-\lambda_0\,{}^t\mu)\}}\,\phi\left((\lambda_0,0,0)\circ g\right), \tag{1.4.11}$$

where $g_0 = (\lambda_0,\mu_0,\kappa_0)$, $g = (\lambda,\mu,\kappa) \in G$ and $\phi \in \mathcal{H}^c$. (1.4.11) can be expressed as follows.

$$\left(U_c(g_0)\phi\right)(g) = e^{2\pi i\sigma\{c(\kappa_0+\kappa+\mu_0\,{}^t\lambda_0+\mu\,{}^t\lambda+2\lambda\,{}^t\mu_0)\}}\,\phi((\lambda_0+\lambda,0,0)). \tag{1.4.12}$$

Theorem 6. *Let c be a positive symmetric half-integral matrix of degree m. Then the Schrödinger representation U_c of G is irreducible.*

Proof. The proof can be found in [42], Theorem 3. □

We let dU_c be the infinitesimal representation associated to the Schrödinger representation U_c. If X is an element of the Lie algebra of G, then

$$dU_c(X)f = \frac{d}{dt}\bigg|_{t=0} U_c(\exp tX)f, \quad f \in \mathcal{H}_c \text{ or } \mathcal{H}^c.$$

We fix an element $\Omega \in \mathbb{H}_n$ once and for all. We let c be a positive symmetric real matrix of degree m. For each $J \in \mathbb{Z}_{\geq 0}^{(m,n)}$, we put

$$f_{c,J}(\xi) := e^{2\pi i\, \sigma(c\xi\Omega\, {}^t\xi)}\, \xi^J, \quad \xi \in \mathbb{R}^{(m,n)}. \tag{1.4.13}$$

Then the set $\left\{ f_{c,J} \mid J \in \mathbb{Z}_{\geq 0}^{(m,n)} \right\}$ forms a basis of $L^2\big(\mathbb{R}^{(m,n)}, d\xi\big) \cong \mathcal{H}_c$.

Proposition 3. *Let $c = (c_{kl})$ be a positive symmetric real matrix of degree m. For each $J \in \mathbb{Z}_{\geq 0}^{(m,n)}$ we have*

$$dU_c(D_{kl}^0)f_{c,J}(\xi) = 2\pi i\, c_{kl}\, f_{c,J}(\xi), \quad 1 \leq k \leq l \leq m, \tag{1.4.14}$$

$$\begin{aligned} dU_c(D_{ka})f_{c,J}(\xi) = 4\pi i \sum_{l=1}^{m}\sum_{b=1}^{n} c_{kl}\Omega_{ab} f_{c,J+\epsilon_{lb}}(\xi) \\ + J_{ka}\, f_{c,J-\epsilon_{ka}}(\xi), \end{aligned} \tag{1.4.15}$$

$$dU_c(\widehat{D}_{lb})f_{c,J}(\xi) = 4\pi i \sum_{p=1}^{m} c_{lp}\, f_{c,J+\epsilon_{pb}}(\xi). \tag{1.4.16}$$

Here $1 \leq k, l \leq m$ and $1 \leq a, b \leq n$.

Proof. We put $E_{kl}^0 = \frac{1}{2}(E_{kl} + E_{lk})$, where $1 \leq k, l \leq m$.

$$\begin{aligned} dU_c(D_{kl}^0)f_{c,J}(\xi) &= \frac{d}{dt}\bigg|_{t=0} U_c\big(\exp tX_{kl}^0\big) f_{c,J}(\xi) \\ &= \frac{d}{dt}\bigg|_{t=0} U_c\big((0,0,tE_{kl}^0)\big) f_{c,J}(\xi) \\ &= \lim_{t\to 0} \frac{e^{2\pi i\sigma(tcE_{kl}^0)} - I}{t} f_{c,J}(\xi) \\ &= \lim_{t\to 0} \frac{e^{2\pi i t c_{kl}} - I}{t} f_{c,J}(\xi) \\ &= 2\pi i\, c_{kl}\, f_{c,J}(\xi). \end{aligned}$$

$$\begin{aligned} dU_c(D_{ka})f_{c,J}(\xi) &= \frac{d}{dt}\bigg|_{t=0} U_c(\exp tX_{ka}) f_{c,J}(\xi) \\ &= \frac{d}{dt}\bigg|_{t=0} U_c((tE_{ka},0,0)) f_{c,J}(\xi) \\ &= \frac{d}{dt}\bigg|_{t=0} e^{2\pi i\, \sigma\{c(\xi+tE_{ka})\Omega\, {}^t(\xi+tE_{ka})\}}\, (\xi + tE_{ka})^J \\ &= 4\pi i \sum_{l=1}^{m}\sum_{b=1}^{n} c_{kl}\Omega_{ab} f_{c,J+\epsilon_{lb}}(\xi) + J_{ka}\, f_{c,J-\epsilon_{ka}}(\xi). \end{aligned}$$

Finally,

$$\begin{aligned} dU_c(\widehat{D}_{lb})f_{c,J}(\xi) &= \frac{d}{dt}\bigg|_{t=0} U_c(\exp t\widehat{X}_{lb})f_{c,J}(\xi) \\ &= \frac{d}{dt}\bigg|_{t=0} U_c((0,tE_{lb},0))f_{c,J}(\xi) \\ &= \lim_{t\to 0} \frac{e^{4\pi i\,\sigma(tc\xi\,{}^tE_{lb})}-I}{t} f_{c,J}(\xi) \\ &= 4\pi i \sum_{p=1}^{m} c_{lp} f_{c,J+\epsilon_{pb}}(\xi). \end{aligned}$$

□

For each $J \in \mathbb{Z}_{\geq 0}^{(m,n)}$, we put

$$\phi_{c,J}(g) = e^{2\pi i\,\sigma\{c(\kappa+\mu\,{}^t\lambda)\}}\, f_{c,J}(\lambda), \tag{1.4.17}$$

where $g = (\lambda, \mu, \kappa) \in G$. Then the set $\{\, \phi_{c,J} \mid J \in \mathbb{Z}_{\geq 0}^{(m,n)} \}$ is a basis of $\mathcal{H}^c$.

Proposition 4. *For each $J \in \mathbb{Z}_{\geq 0}^{(m,n)}$ and $g = (\lambda, \mu, \kappa) \in G$, we have*

$$dU_c(D_{kl}^0)\,\phi_{c,J}(g) = 2\pi i\, c_{kl}\,\phi_{c,J}(g), \quad 1 \leq k \leq l \leq m, \tag{1.4.18}$$

$$\begin{aligned} dU_c(D_{ka})\phi_{c,J}(g) &= 4\pi i \sum_{l=1}^{m}\sum_{b=1}^{n} c_{kl}\,\Omega_{ab}\,\phi_{c,J+\epsilon_{lb}}(g) \\ &\quad + J_{ka}\,\phi_{c,J-\epsilon_{ka}}(g), \end{aligned} \tag{1.4.19}$$

$$dU_c(\widehat{U}_{lb})\phi_{c,J}(g) = 2\pi i \sum_{p=1}^{m} c_{lp}\phi_{c,J+\epsilon_{pb}}(g). \tag{1.4.20}$$

Here $1 \leq k,l \leq m$ and $1 \leq a,b \leq n$.

Proof. We put $E_{kl}^0 = \frac{1}{2}(E_{kl} + E_{lk})$, where $1 \leq k,l \leq m$. Then we have

$$\begin{aligned} dU_c(D_{kl}^0)\phi_{c,J}(g) &= \frac{d}{dt}\bigg|_{t=0} U_c(\exp tX_{kl}^0)\phi_{c,J}(g) \\ &= \frac{d}{dt}\bigg|_{t=0} U_c((0,0,tE_{kl}^0))\phi_{c,J}(g) \\ &= \lim_{t\to 0} \frac{e^{2\pi i\,\sigma(tcE_{kl}^0)}-I}{t}\phi_{c,J}(g) \\ &= 2\pi i\, c_{kl}\,\phi_{c,J}(g). \end{aligned}$$

And we have

$$\begin{aligned}
dU_c(D_{ka})\phi_{c,J}(g) &= \frac{d}{dt}\bigg|_{t=0} U_c(\exp tX_{ka})\phi_{c,J}(g) \\
&= \frac{d}{dt}\bigg|_{t=0} U_c((tE_{ka},0,0))\phi_{c,J}(g) \\
&= \frac{d}{dt}\bigg|_{t=0} e^{-2\pi it\,\sigma(cE_{ka}\,{}^t\mu)}\,\phi_{c,J}((tE_{ka},0,0)\circ g) \\
&= \frac{d}{dt}\bigg|_{t=0} e^{-2\pi it\,\sigma(cE_{ka}\,{}^t\mu)}\cdot e^{2\pi i\,\sigma\{c(\kappa+tE_{ka}{}^t\mu+\mu\,{}^t\lambda+t\mu\,{}^tE_{ka})\}} \\
&\quad\times e^{2\pi i\,\sigma\{c(\lambda+tE_{ka})\Omega\,{}^t(\lambda+tE_{ka})\}}\,(\lambda+tE_{ka})^J \\
&= e^{2\pi i\,\sigma\{c(\kappa+\mu\,{}^t\lambda+\lambda\Omega\,{}^t\lambda)\}} \\
&\quad\times \frac{d}{dt}\bigg|_{t=0} e^{4\pi it\,\sigma(c\lambda\Omega\,{}^tE_{ka})+2\pi it^2\,\sigma(cE_{ka}\Omega\,{}^tE_{ka})}\,(\lambda+tE_{ka})^J \\
&= 4\pi i\sum_{l=1}^{m}\sum_{b=1}^{n} c_{kl}\,\Omega_{ab}\,\phi_{c,J+\epsilon_{lb}}(g) + J_{ka}\,\phi_{c,J-\epsilon_{ka}}(g).
\end{aligned}$$

Finally,

$$\begin{aligned}
dU_{\widehat{D}_{lb}}(\sigma_c)\,\phi_{c,J}(g) &= \frac{d}{dt}\bigg|_{t=0} U_c(\exp t\widehat{X}_{lb})\phi_{c,J}(g) \\
&= \frac{d}{dt}\bigg|_{t=0} U_c((0,tE_{lb},0))\phi_{c,J}(g) \\
&= \frac{d}{dt}\bigg|_{t=0} e^{2\pi it\,\sigma(c\lambda\,{}^tE_{lb})}\phi_{c,J}(g) \\
&= \lim_{t\to 0}\frac{e^{2\pi it(\sum_{p=1}^m c_{lp}\lambda_{pb})}-I}{t}\,\phi_{c,J}(g) \\
&= 2\pi i\left(\sum_{p=1}^{m} c_{lp}\lambda_{pb}\right)\phi_{c,J}(g) \\
&= 2\pi i\sum_{p=1}^{m} c_{lp}\phi_{c,J+\epsilon_{pb}}(g).
\end{aligned}$$

□

1.5 Fock Representations

We consider the vector space $V^{(m,n)} := \mathbb{R}^{(m,n)} \times \mathbb{R}^{(m,n)}$. We put

$$P_{ka} = (E_{ka}, 0), \quad Q_{lb} = (0, E_{lb}), \tag{1.5.1}$$

where $1 \le k, l \le m$ and $1 \le a, b \le n$. Then the set $\{P_{ka},\, Q_{ka}\}$ forms a basis for $V^{(m,n)}$. We define the alternating bilinear form $\mathbf{A} : V^{(m,n)} \times V^{(m,n)} \longrightarrow \mathbb{R}$ by

$$\mathbf{A}\left((\lambda_0, \mu_0), (\lambda, \mu)\right) = \sigma(\lambda_0\, {}^t\mu - \mu_0\, {}^t\lambda), \quad (\lambda_0, \mu_0), (\lambda, \mu) \in V^{(m,n)}. \tag{1.5.2}$$

Then we have

$$\mathbf{A}(P_{ka}, P_{lb}) = \mathbf{A}(Q_{ka}, Q_{lb}) = 0, \quad \mathbf{A}(P_{ka}, Q_{lb}) = \delta_{ab}\, \delta_{kl}, \tag{1.5.3}$$

where $1 \le k, l \le m$ and $1 \le a, b \le n$. Any element $v \in V^{(m,n)}$ can be written uniquely as

$$v = \sum_{k,a} x_{ka} P_{ka} + \sum_{l,b} y_{lb} Q_{lb}, \quad x_{ka},\, y_{lb} \in \mathbb{R}. \tag{1.5.4}$$

From now on, for brevity, we write $V := V^{(m,n)}$ and $v = xP + yQ$ instead of (1.5.4). Then it is easy to see that the endomorphism $J : V \longrightarrow V$ defined by

$$J(xP + yQ) := -yP + xQ, \quad xP + yQ \in V \tag{1.5.5}$$

is a complex structure on V which is compatible with the alternating bilinear form $\mathbf{A}$. This means that J is an endomorphism of V satisfying the following conditions:

(J1) $J^2 = -I$ on V.

(J2) $\mathbf{A}(Jv_0, Jv) = \mathbf{A}(v_0, v)$ for all $v_0, v \in V$.

(J3) $\mathbf{A}(v, Jv) > 0$ for all $v \in V$ with $v \neq 0$.

Now we let $V_{\mathbb{C}} = V + i\,V$ be the complexification of V, where $i = \sqrt{-1}$. For an element $w = v_1 + i\,v_2 \in V_{\mathbb{C}}$ with $v_1, v_2 \in V$, we put

$$\overline{w} := v_1 - i\,v_2. \tag{1.5.6}$$

Let $\mathbf{A}_{\mathbb{C}}$ be the complex bilinear form on $V_{\mathbb{C}}$ extending $\mathbf{A}$ and let $J_{\mathbb{C}}$ be the complex linear map of $V_{\mathbb{C}}$ extending J. Since $J_{\mathbb{C}}^2 = -I$, $J_{\mathbb{C}}$ has the only eigenvalues $\pm\, i$. We denote by V^+ (resp. V^-) the eigenspace of $V_{\mathbb{C}}$ corresponding to the eigenvalues i (resp. $-i$). Thus $V_{\mathbb{C}} = V^+ + V^-$. Since

$$J_{\mathbb{C}}(P_{ka} \pm i\,Q_{ka}) = \mp i\,(P_{ka} \pm i\,Q_{ka}),$$

we have

$$V^+ = \sum_{k,a} \mathbb{C}\,(P_{ka} - i\,Q_{ka}), \quad V^- = \sum_{k,a} \mathbb{C}\,(P_{ka} + i\,Q_{ka}). \tag{1.5.7}$$

Let

$$V_* := \sum_{k,a} \mathbb{C}\,P_{ka}, \quad 1 \le k \le m, \ \ 1 \le a \le n \tag{1.5.8}$$

be the subspace of $V_{\mathbb{C}}$ as a $\mathbb{C}$-vector space. It is easy to see that V_* is isomorphic to V as $\mathbb{R}$-vector spaces via the isomorphism $T : V \longrightarrow V_*$ defined by

$$T(P_{ka}) = P_{ka}, \quad T(Q_{lb}) = i\,P_{lb}. \tag{1.5.9}$$

We define the complex linear map $J_* : V_* \longrightarrow V_*$ by $J_*(P_{ka}) = i\,P_{ka}$ for $1 \le k \le m$, $1 \le a \le n$. Then J_* is compatible with J, that is, $T \circ J = J_* \circ T$. It is easily seen that there exists a unique hermitian form $\mathbf{H}$ on V_* with $\operatorname{Im}\mathbf{H} = \mathbf{A}$. Indeed, $\mathbf{H}$ is given by

$$\mathbf{H}(v,w) = \mathbf{A}(v, J_* w) + i\,\mathbf{A}(v,w), \quad v, w \in V_*. \tag{1.5.10}$$

For $v = \sum_{k,a} z_{ka} P_{ka} \in V_*$ with $z_{ka} = x_{ka} + i y_{ka}$ $(x_{ka}, y_{ka} \in \mathbb{R})$, for brevity we write $v = zP$. For two elements $v = zP$ and $v' = z'P$ in V_*, $\mathbf{H}(v,v') = \sum_{k,a} \overline{z_{ka}}\, z'_{ka}$. We observe that

$$V_{\mathbb{C}} = \sum_{k,a} \mathbb{C}\,P_{ka} + \sum_{l,b} \mathbb{C}\,Q_{lb} = V^+ + V^- \supset V^{\pm}.$$

For $w = z^0 P + z^1 Q \in V_{\mathbb{C}}$, we put

$$w = w^+ + w^-, \quad w^+ := z^+(P - i\,Q), \quad w^- := z^-(P + i\,Q).$$

The relations among z^0, z^1, z^+, z^- are given by

$$z^{\pm} = \frac{1}{2}(z^0 \pm i\,z^1), \quad z^0 = z^+ + z^-, \quad z^1 = i\,(z^- - z^+). \tag{1.5.11}$$

Precisely, (1.5.11) implies that

$$z^{\pm}_{ka} = \frac{1}{2}\,(z^0_{ka} \pm i\,z^1_{ka}), \quad z^0_{ka} = z^+_{ka} + z^-_{ka}, \quad z^1_{ka} = i\,(z^-_{ka} - z^+_{ka}),$$

where $1 \le k \le m$ and $1 \le a \le n$. It is easy to see that

$$\mathbf{A}_{\mathbb{C}}(w^-, w^+) = -2i \sum_{k,a} z^-_{ka} z^+_{ka} = -\frac{i}{2} \sum_{k,a} \left\{ (z^0_{ka})^2 + (z^1_{ka})^2 \right\}. \tag{1.5.12}$$

Let

$$G_{\mathbb{C}} := \left\{ (z^0, z^1, a) \mid z^0, z^1 \in \mathbb{C}, \ \ a \in \mathbb{C}^{(m,m)}, \ \ a + z^1\, {}^t z^0 \text{ symmetric} \right\}$$

be the complexification of the real Heisenberg group $G := H_{\mathbb{R}}^{(n,m)}$. Analogously in the real case, the multiplication on $G_{\mathbb{C}}$ is given by (1.1.1). If $w = z^0 P + z^1 Q := \sum_{k,a} z^0_{ka} P_{ka} + \sum_{l,b} z^1_{lb} Q_{lb}$, we identify z^0, z^1 with the $m \times n$ matrices respectively:

$$z^0 := \begin{pmatrix} z^0_{11} & z^0_{12} & \cdots & z^0_{1n} \\ z^0_{21} & z^0_{22} & \cdots & z^0_{2n} \\ \vdots & \vdots & \ddots & \vdots \\ z^0_{m1} & z^0_{m2} & \cdots & z^0_{mn} \end{pmatrix}, \quad z^1 := \begin{pmatrix} z^1_{11} & z^1_{12} & \cdots & z^1_{1n} \\ z^1_{21} & z^1_{22} & \cdots & z^1_{2n} \\ \vdots & \vdots & \ddots & \vdots \\ z^1_{m1} & z^1_{m2} & \cdots & z^1_{mn} \end{pmatrix}.$$

That is, we identify $w = z^0 P + z^1 Q \in V_{\mathbb{C}}$ with $(z^0, z^1) \in \mathbb{C}^{(m,n)} \times \mathbb{C}^{(m,n)}$. If $w = z^0 P + z^1 Q,\ \hat{w} = \hat{z}^0 P + \hat{z}^1 Q \in V_{\mathbb{C}}$, then

$$(w,a) \circ (\hat{w}, \hat{a}) = (w + \hat{w}, a + \hat{a} + z^0\, {}^t\hat{z}^1 - z^1\, {}^t\hat{z}^0), \quad a, \hat{a} \in \mathbb{C}^{(m,m)}. \tag{1.5.13}$$

From now on, for brevity we put

$$R^+ := P - i\,Q, \qquad R^- := P + i\,Q. \tag{1.5.14}$$

If $w = z^+ R^+ + z^- R^-,\ \hat{w} = \hat{z}^+ R^+ + \hat{z}^- R^- \in V_{\mathbb{C}}$, by an easy computation, we have

$$(w,a) \circ (\hat{w}, \hat{a}) = (\tilde{w}, a + \hat{a} + 2i\,(z^+\, {}^t\hat{z}^- - z^-\, {}^t\hat{z}^+)) \tag{1.5.15}$$

with

$$\tilde{w} = (z^+ + \hat{z}^+) R^+ + (z^- + \hat{z}^-) R^-.$$

Here we identified z^+, z^- with $m \times n$ matrices

$$z^+ := \begin{pmatrix} z^+_{11} & z^+_{12} & \cdots & z^+_{1n} \\ z^+_{21} & z^+_{22} & \cdots & z^+_{2n} \\ \vdots & \vdots & \ddots & \vdots \\ z^+_{m1} & z^+_{m2} & \cdots & z^+_{mn} \end{pmatrix}, \quad z^- := \begin{pmatrix} z^-_{11} & z^-_{12} & \cdots & z^-_{1n} \\ z^-_{21} & z^-_{22} & \cdots & z^-_{2n} \\ \vdots & \vdots & \ddots & \vdots \\ z^-_{m1} & z^-_{m2} & \cdots & z^-_{mn} \end{pmatrix}.$$

It is easy to see that

$$P_{\mathbb{C}} := \left\{ (w^-, a) \in G_{\mathbb{C}} \mid w^- \in V^-, \ \ a \in \mathbb{C}^{(m,m)} \right\} \tag{1.5.16}$$

is a commutative subgroup of $G_{\mathbb{C}}$ and

$$G \cap P_{\mathbb{C}} = \mathcal{Z}, \quad G_{\mathbb{C}} = G \circ P_{\mathbb{C}},$$

where $\mathcal{Z} := \left\{ (0,0,\kappa) \in G \mid \kappa = {}^t\kappa \in \mathbb{R}^{(m,m)} \right\} \cong \mathrm{Sym}(m, \mathbb{R})$ is the center of G. Moreover,

$$P_{\mathbb{C}} \backslash G_{\mathbb{C}} \cong V^+ \cong \mathbb{R}^{(m,n)} \times \mathbb{R}^{(m,n)} \cong \mathcal{Z} \backslash G. \tag{1.5.17}$$

For $c = {}^t c \in \mathrm{Sym}(m, \mathbb{R})$ with $c > 0$, we let $\delta_c : P_{\mathbb{C}} \longrightarrow \mathbb{C}^\times$ be a quasi-character of $P_{\mathbb{C}}$ defined by

$$\delta_c((w^-, a)) = e^{2\pi i \sigma(ca)}, \quad (w^-, a) \in P_{\mathbb{C}}. \tag{1.5.18}$$

Let

$$U^{F,c} = \mathrm{Ind}_{P_{\mathbb{C}}}^{G_{\mathbb{C}}} \delta_c$$

be the representation of $G_{\mathbb{C}}$ induced from a quasi-character δ_c of $P_{\mathbb{C}}$. Then $U^{F,c}$ is realized in the Hilbert space $\mathcal{H}^{F,c}$ consisting of all holomorphic functions $\psi : G_{\mathbb{C}} \longrightarrow \mathbb{C}$ satisfying the following conditions:

(F1) $\psi((w^-, a) \circ g) = \delta_c((w^-, a))\psi(g) = e^{2\pi i\, \sigma(ca)}\, \psi(g)$ for all $(w^-, a) \in P_{\mathbb{C}}$ and $g \in G_{\mathbb{C}}$.

(F2) $\int_{\mathcal{Z}\backslash G} |\psi(\dot{g})|^2 \, d\dot{g} < \infty$.

The inner product $\langle \ , \ \rangle_{F,c}$ on $\mathcal{H}^{F,c}$ is given by

$$\langle \psi_1, \psi_2 \rangle_{F,c} := \int_{\mathcal{Z}\backslash G} \psi_1(\dot{g})\, \overline{\psi_2(\dot{g})}\, d\dot{g}, \quad \psi_1, \psi_2 \in \mathcal{H}^{F,c}, \ \dot{g} = \mathcal{Z}g.$$

$U^{F,c}$ is realized by the right regular representation of $G_{\mathbb{C}}$ on $\mathcal{H}^{F,c}$:

$$\left(U^{F,c}(g_0)\psi\right)(g) = \psi(gg_0), \quad \psi \in \mathcal{H}^{F,c}, \ g_0, g \in G_{\mathbb{C}}. \tag{1.5.19}$$

Now we will show that $U^{F,c}$ is realized as a representation of G in the Fock space. The Fock space $\mathcal{H}_{F,c}$ is the Hilbert space consisting of all holomorphic functions $f : \mathbb{C}^{(m,n)} \cong V_* \longrightarrow \mathbb{C}$ satisfying the condition

$$\| f \|_{F,c}^2 = \int_{\mathbb{C}^{(m,n)}} |f(W)|^2 \, e^{-2\pi\sigma(c\, W\, {}^t\overline{W})} \, dW < \infty.$$

The inner product $(\ , \ )_{F,c}$ on $\mathcal{H}_{F,c}$ is given by

$$(f_1, f_2)_{F,c} = \int_{\mathbb{C}^{(m,n)}} f_1(W)\, \overline{f_2(W)}\, e^{-2\pi\sigma(c\, W\, {}^t\overline{W})} \, dW, \quad f_1, f_2 \in \mathcal{H}_{F,c}.$$

Lemma 6. *The mapping* $\Lambda : \mathcal{H}_{F,c} \longrightarrow \mathcal{H}^{F,c}$, $\Lambda_f := \Lambda(f)\, (f \in \mathcal{H}_{F,c})$ *defined by*

$$\Lambda_f((z^0 P + z^1 Q, a)) = e^{2\pi i\, \sigma\{c(a + 2\, i\, z^-\, {}^t z^+)\}} \, f(2z^+) \tag{1.5.20}$$

is an isometry of $\mathcal{H}_{F,c}$ *onto* $\mathcal{H}^{F,c}$, *where* $2z^\pm = z^0 \pm i\, z^1$ *(cf. (6.11)). The inverse* $\Delta : \mathcal{H}^{F,c} \longrightarrow \mathcal{H}_{F,c}$, $\Delta_\psi := \Delta(\psi)\, (\psi \in \mathcal{H}^{F,c})$ *is given by*

$$\Delta_\psi(W) = \psi\left(\frac{1}{2} W R^+\right), \quad W \in \mathbb{C}^{(m,n)}, \tag{1.5.21}$$

where $R^{\pm} = P \mp i\,Q$ *(cf. (1.5.14)).*

Proof. First we observe that for $w = z^0 P + z^1 Q = z^+ R^+ + z^- R^- \in V_{\mathbb{C}}$,

$$(w,a) = (z^- R^-, a + 2\,i\,z^-\,{}^t z^+) \circ (z^+ R^+, 0).$$

Thus if $\psi \in \mathcal{H}^{F,c}$ and $w = z^0 P + z^1 Q = z^+ R^+ + z^- R^-$, by (F1),

$$\psi((w,a)) = e^{2\pi i\,\sigma\{c(a + 2\,i\,z^-\,{}^t z^+)\}}\,\psi((z^+ R^+, 0)). \tag{1.5.22}$$

Let $W = x + i\,y \in \mathbb{C}^{(m,n)}$ with $x, y \in \mathbb{R}^{(m,n)}$. Then

$$xP + yQ = z^+ R^+ + z^- R^-, \quad 2z^{\pm} = x \pm i\,y.$$

So $z^-\,{}^t z^+ = \frac{1}{4}\,W\,{}^t\overline{W}$. According to (1.5.22), if $\psi \in \mathcal{H}^{F,c}$, we have

$$\psi((xP + yQ, 0)) = e^{-\pi\,\sigma(c\,W\,{}^t\overline{W})}\,\psi\left(\left(\frac{1}{2}WR^+, 0\right)\right).$$

Thus we get

$$|\psi((xP + yQ, 0))|^2 = e^{-2\pi\,\sigma(c\,W\,{}^t\overline{W})}\left|\psi\left(\left(\frac{1}{2}WR^+, 0\right)\right)\right|^2.$$

Therefore

$$\int_{\mathcal{Z}\backslash G} |\psi(\dot{g})|\,d\dot{g} = \int_{\mathbb{C}^{(m,n)}} e^{-2\pi\sigma(c\,W\,{}^t\overline{W})}\,|\Delta_\psi(W)|^2 dW < \infty.$$

It is easy to see that Δ is the inverse of Λ. Hence we obtain the desired results. □

Lemma 7. *The representation* $U^{F,c}$ *is realized as a representation of* G *in the Fock space* $\mathcal{H}_{F,c}$ *as follows. If* $g = (\lambda P + \mu Q, \kappa) = (\lambda, \mu, \kappa) \in G$ *and* $f \in \mathcal{H}_{F,c}$, *then*

$$\left(U^{F,c}(g)f\right)(W) = e^{2\pi i\,\sigma(c\,\kappa)}\,e^{-\pi\,\sigma\{c\,(\zeta\,{}^t\bar{\zeta} + 2\,W\,{}^t\bar{\zeta})\}}\,f(W + \zeta), \tag{1.5.23}$$

where $W \in \mathbb{C}^{(m,n)}$ *and* $\zeta = \lambda + i\,\mu$.

Proof.

$$\begin{aligned}
\left(U^{F,c}(g)f\right)(W) &= \left(\Delta(U^{F,c}(g)(\Lambda_f))\right)(W)\\
&= \left(U^{F,c}(g)(\Lambda_f)\right)\left(\frac{1}{2}WR^+\right)\\
&= \Lambda_f\left(\left(\frac{1}{2}WR^+,0\right)\circ g\right)\\
&= \Lambda_f\left(\left(\frac{1}{2}W,-\frac{i}{2}W,0\right)\circ(\lambda,\mu,\kappa)\right)\\
&= \Lambda_f\left(\left(\lambda+\frac{1}{2}W\right)P+\left(\mu-\frac{i}{2}W\right)Q,\kappa+\frac{1}{2}W\,{}^t\mu+\frac{i}{2}W\,{}^t\lambda\right)\\
&= e^{2\pi i\,\sigma\{c(\kappa+\frac{i}{2}W\,{}^t\overline{\zeta}+\frac{i}{2}\overline{\zeta}\,{}^tW+\frac{i}{2}\overline{\zeta}\,{}^t\zeta)\}}\,f(W+\zeta)\qquad(**)\\
&= e^{2\pi i\,\sigma(c\kappa)}\cdot e^{-\pi\,\sigma\{\mathcal{M}(\zeta\,{}^t\overline{\zeta}+W\,{}^t\overline{\zeta})\}}\,f(W+\zeta),
\end{aligned}$$

where $\zeta=\lambda+i\mu$. In (**), we used (1.5.20) and the facts that $2iz^-\,{}^tz^+=\frac{i}{2}(\overline{W}\,{}^t\zeta+\overline{W}\,{}^tW)$ and $2z^+=W+\zeta$. □

Definition 4. The induced representation $U^{F,c}$ of G in the Fock space $\mathcal{H}_{F,c}$ is called the **Fock representation** of G.

Let $W=U+iV\in\mathbb{C}^{(m,n)}$ with $U,V\in\mathbb{R}^{(m,n)}$. If $U=(u_{ka})$, $V=(v_{lb})$ are coordinates in $\mathbb{C}^{(m,n)}$, we put

$$dU=du_{11}du_{12}\cdots du_{mn},\quad dV=dv_{11}dv_{12}\cdots dv_{mn}$$

and $dW=dUdV$. And we set

$$d\mu(W)=e^{-\pi\sigma(W\,{}^t\overline{W})}\,dW. \tag{1.5.24}$$

Let f be a holomorphic function on $\mathbb{C}^{(m,n)}$. Then $f(W)$ has the Taylor expansion

$$f(z)=\sum_{J\in\mathbb{Z}_{\geq 0}^{(m,n)}}a_JW^J,\quad W=(w_{ka})\in\mathbb{C}^{(m,n)},$$

where $J=(J_{ka})\in J\in\mathbb{Z}_{\geq 0}^{(m,n)}$ and $W^J:=w_{11}^{J_{11}}w_{12}^{J_{12}}\cdots w_{mn}^{J_{mn}}$.

We set $|W|_\infty:=\max_{k,a}(|w_{ka}|)$. Then by an easy computation, we have

$$\begin{aligned}
\int_{\mathbb{C}^{(m,n)}}|f(W)|^2\,d\mu(W) &= \lim_{r\to\infty}\int_{|W|_\infty\leq r}|f(W)|^2d\mu(W)\\
&= \lim_{r\to\infty}\sum_{J,K}a_J\overline{a_K}\int_{|W|_\infty\leq r}W^J\overline{W^K}\,d\mu(W)\\
&= \sum_J|a_J|^2\pi^{-|J|}J!,
\end{aligned}$$

where J runs over $J\in\mathbb{Z}_{\geq 0}^{(m,n)}$.

Let $\mathcal{H}_{m,n}$ be the Hilbert space consisting of all holomorphic functions $f : \mathbb{C}^{(m,n)} \longrightarrow \mathbb{C}$ satisfying the condition

$$\|f\|^2 = \int_{\mathbb{C}^{(m,n)}} |f(W)|^2 \, d\mu(W) < \infty. \tag{1.5.25}$$

The inner product (,) on $\mathcal{H}_{m,n}$ is given by

$$(f_1, f_2) = \int_{\mathbb{C}^{(m,n)}} f_1(W)\, \overline{f_2(W)}\, d\mu(W), \quad f_1, f_2 \in \mathcal{H}_{m,n}.$$

Thus we have

Lemma 8. *Let $f \in \mathcal{H}_{m,n}$ and let $f(W) = \sum_J a_J W^J$ be the Taylor expansion of f. Then*

$$\|f\|^2 = \sum_{J \in \mathbb{Z}_{\geq 0}^{(m,n)}} |a_J|^2 \pi^{-|J|} J!.$$

For each $J \in \mathbb{Z}_{\geq 0}^{(m,n)}$, we define the holomorphic function $\Phi_J(W)$ on $\mathbb{C}^{(m,n)}$ by

$$\Phi_J(W) := (J!)^{-\frac{1}{2}} \left(\pi^{\frac{1}{2}} W\right)^J, \quad W \in \mathbb{C}^{(m,n)}. \tag{1.5.26}$$

Then

$$(\Phi_J, \Phi_K) = \begin{cases} 1 & \text{if } J = K \\ 0 & \text{otherwise.} \end{cases} \tag{1.5.27}$$

It is easy to see that the set $\left\{ \Phi_J \mid J \in \mathbb{Z}_{\geq 0}^{(m,n)} \right\}$ forms a complete orthonormal system in $\mathcal{H}_{m,n}$. By the Schwarz inequality, for any $f \in \mathcal{H}_{m,n}$, we have

$$|f(W)| \leq e^{\frac{\pi}{2}\sigma(W\, {}^t\overline{W})}\, \|f\|, \quad W \in \mathbb{C}^{(m,n)}. \tag{1.5.28}$$

Consequently, the norm convergence in $\mathcal{H}_{m,n}$ implies the uniform convergence on any bounded subset of $\mathbb{C}^{(m,n)}$. We observe that for a fixed $W' \in \mathbb{C}^{(m,n)}$, the holomorphic function $W \longrightarrow e^{\pi\sigma(W\, {}^t\overline{W'})}$ admits the following Taylor expansion

$$e^{\pi\sigma(W\, {}^t\overline{W'})} = \sum_{J \in \mathbb{Z}_{\geq 0}^{(m,n)}} \Phi_J(W)\, \Phi_J(\overline{W'}). \tag{1.5.29}$$

From (1.5.29), we obtain

$$\Phi_J(\overline{W'}) = (J!)^{-\frac{1}{2}} \int_{\mathbb{C}^{(m,n)}} e^{\pi\sigma(W\, {}^t\overline{W'})} \left(\pi^{\frac{1}{2}} \overline{W}\right)^J d\mu(W). \tag{1.5.30}$$

Thus if $f \in \mathcal{H}_{m,n}$, we get

$$\left(f(W),\, e^{\pi\sigma(W\,{}^t\overline{W'})}\right) = \left(f,\, \sum_J \Phi_J(\overline{W'})\,\Phi_J(\cdot)\right)$$
$$= \sum_J \Phi_J(W')\,(f,\Phi_J)$$
$$= f(W').$$

Hence $e^{\pi\sigma(W\,{}^t\overline{W'})}$ is the reproducing kernel for $\mathcal{H}_{m,n}$ in the sense that for any $f \in \mathcal{H}_{m,n}$,

$$f(W) = \int_{\mathbb{C}^{(m,n)}} e^{\pi\sigma(W\,{}^t\overline{W'})}\, f(W')\, d\mu(W'). \tag{1.5.31}$$

We set

$$\kappa(W,W') := e^{\pi\sigma(W\,{}^t\overline{W'})}, \quad W, W' \in \mathbb{C}^{(m,n)}. \tag{1.5.32}$$

Obviously $\kappa(W,W') = \overline{\kappa(W',W)}$. (1.5.31) may be written as

$$f(W) = \int_{\mathbb{C}^{(m,n)}} \kappa(W,W')\, f(W')\, d\mu(W'), \quad f \in \mathcal{H}_{m,n}. \tag{1.5.33}$$

Let $\mathcal{M}$ be a positive definite, symmetric half-integral matrix of degree m. We define the measure

$$d\mu_{\mathcal{M}}(W) = e^{-2\pi\,\sigma(\mathcal{M}W\,{}^t\overline{W})}\, dW. \tag{1.5.34}$$

We recall the Fock space $\mathcal{H}_{F,\mathcal{M}}$ consisting of all holomorphic functions $f : \mathbb{C}^{(m,n)} \longrightarrow \mathbb{C}$ that satisfy the condition

$$\|f\|^2_{\mathcal{M}} := \|f\|^2_{F,\mathcal{M}} := \int_{\mathbb{C}^{(m,n)}} |f(W)|^2\, d\mu_{\mathcal{M}}(W) < \infty. \tag{1.5.35}$$

The inner product $(\ ,\ )_{\mathcal{M}} := (\ ,\ )_{F,\mathcal{M}}$ on $\mathcal{H}_{F,\mathcal{M}}$ is given by

$$(f_1, f_2)_{\mathcal{M}} = \int_{\mathbb{C}^{(m,n)}} f_1(W)\,\overline{f_2(W)}\, d\mu_{\mathcal{M}}(W), \quad f_1, f_2 \in \mathcal{H}_{F,\mathcal{M}}.$$

Lemma 9. *Let $f \in \mathcal{H}_{F,\mathcal{M}}$ and let $g(W) = f\left((2\mathcal{M})^{-\frac{1}{2}}W\right)$ be the holomorphic function on $\mathbb{C}^{(m,n)}$. We let*

$$g(W) = \sum_{J \in \mathbb{Z}^{(m,n)}_{\geq 0}} a_{\mathcal{M},J}\, W^J$$

be the Taylor expansion of $g(W)$. Then we have

$$\|f\|^2_{\mathcal{M}} = (f,f)_{\mathcal{M}} = 2^{-n}(\det \mathcal{M})^{-n} \sum_{J \in \mathbb{Z}^{(m,n)}_{\geq 0}} |a_{\mathcal{M},J}|^2\, \pi^{-|J|}\, J!.$$

Proof. Let $\mathcal{M}^{\frac{1}{2}}$ be the unique positive definite symmetric matrix of degree m such that $\left(\mathcal{M}^{\frac{1}{2}}\right)^2 = \mathcal{M}$. We put $\widetilde{W} := \sqrt{2}\mathcal{M}^{\frac{1}{2}}W$. Obviously $d\widetilde{W} = 2^n\,(det\,\mathcal{M})^n dW$. Thus for $f \in \mathcal{H}_{F,\mathcal{M}}$, we have

$$\begin{aligned}(f,f)_{\mathcal{M}} &= \int_{\mathbb{C}^{(m,n)}} |f(W)|^2\, d\mu_{\mathcal{M}}(W)\\ &= 2^{-n}(\det \mathcal{M})^{-n} \int_{\mathbb{C}^{(m,n)}} |g(W)|^2\, d\mu(W)\\ &= 2^{-n}(\det \mathcal{M})^{-n} \sum_{J\in \mathbb{Z}_{\geq 0}^{(m,n)}} |a_{\mathcal{M},J}|^2 \pi^{-|J|} J! \quad \text{(by Lemma 8)}\end{aligned}$$

□

For each $J \in \mathbb{Z}_{\geq 0}^{(m,n)}$, we put

$$\Phi_{\mathcal{M},J}(W) := 2^{\frac{n}{2}}\,(\det \mathcal{M})^{\frac{n}{2}}\,(J!)^{-\frac{1}{2}}\left((2\pi\mathcal{M})^{\frac{1}{2}}W\right)^J, \quad W \in \mathbb{C}^{(m,n)}. \tag{1.5.36}$$

Lemma 10. *The set* $\left\{ \Phi_{\mathcal{M},J} \,\middle|\, J \in \mathbb{Z}_{\geq 0}^{(m,n)} \right\}$ *is a complete orthonormal system in* $\mathcal{H}_{F,\mathcal{M}}$.

Proof. For $J, K \in \mathbb{Z}_{\geq 0}^{(m,n)}$, we have

$$\begin{aligned}(\Phi_{\mathcal{M},J},\Phi_{\mathcal{M},K})_{\mathcal{M}} &= 2^n(\det \mathcal{M})^n (J!)^{-\frac{1}{2}}(K!)^{-\frac{1}{2}}\\ &\quad\times \int_{\mathbb{C}^{(m,n)}} \left((2\pi\mathcal{M})^{\frac{1}{2}}W\right)^J \left((2\pi\mathcal{M})^{\frac{1}{2}}\overline{W}\right)^K d\mu_{\mathcal{M}}(W)\\ &= (J!)^{-\frac{1}{2}}(K!)^{-\frac{1}{2}} \int_{\mathbb{C}^{(m,n)}} (\pi^{\frac{1}{2}}W)^J\, \overline{(\pi^{\frac{1}{2}}W)^K}\, d\mu(W)\\ &= (\Phi_J, \Phi_K).\end{aligned}$$

By (1.5.27), we have

$$(\Phi_{\mathcal{M},J},\Phi_{\mathcal{M},K})_{\mathcal{M}} = \begin{cases} 1 & \text{if } J = K\\ 0 & \text{otherwise.}\end{cases} \tag{1.5.37}$$

We leave the proof of the completeness to the reader. □

We observe that for a fixed $W' \in \mathbb{C}^{(m,n)}$, the holomorphic function $W \longrightarrow e^{\pi\sigma(\mathcal{M}W\,{}^t\overline{W'})}$ admits the following Taylor expansion

$$e^{\pi\sigma(\mathcal{M}W\,{}^t\overline{W'})} = \sum_{J\in \mathbb{Z}_{\geq 0}^{(m,n)}} \Phi_{\mathcal{M},J}(W)\,\Phi_{\mathcal{M},J}(\overline{W'}). \tag{1.5.38}$$

If $f \in \mathcal{H}_{F,\mathcal{M}}$, we have

$$\left(f(W),\, e^{\pi\sigma(\mathcal{M}W\,{}^t\overline{W'})}\right)_{\mathcal{M}} = \sum_{J\in\mathbb{Z}_{\geq 0}^{(m,n)}} (f,\Phi_{\mathcal{M},J})_{\mathcal{M}}\,\Phi_{\mathcal{M},J}(W') \\ = f(W').$$

Hence $e^{\pi\sigma(\mathcal{M}W\,{}^t\overline{W'})}$ is the reproducing kernel for $\mathcal{H}_{F,\mathcal{M}}$ in the sense that

$$f(W) = \int_{\mathbb{C}^{(m,n)}} f(W')\, e^{\pi\,\sigma(\mathcal{M}W\,{}^t\overline{W'})}\, d\mu_{\mathcal{M}}(W'). \tag{1.5.39}$$

For $U \in \mathbb{R}^{(m,n)}$ and $W \in \mathbb{C}^{(m,n)}$, we put

$$k(U,W) := e^{2\pi\,\sigma(-U\,{}^tU + \frac{1}{2}W\,{}^tW + 2\,i\,U\,{}^tW)}. \tag{1.5.40}$$

Then we have the following lemma.

Lemma 11.

$$\int_{\mathbb{R}^{(m,n)}} k(U,W)\,\overline{k(U,W')}\,dU = e^{2\pi\sigma(W\,{}^tW')}.$$

Proof. We put

$$\mathcal{I}(W,W') := \int_{\mathbb{R}^{(m,n)}} k(U,W)\,\overline{k(U,W')}\,dU.$$

Then we have

$$\mathcal{I}(W,W') = e^{\pi\,\sigma(W\,{}^tW + \overline{W'}\,{}^t\overline{W'})} \int_{\mathbb{R}^{(m,n)}} e^{-4\pi\,\sigma(U\,{}^tU)}\, e^{4\pi i\sigma\{U\,{}^t(W-\overline{W'})\}}\,dU \\ = e^{\pi\,\sigma(W\,{}^tW + \overline{W'}\,{}^t\overline{W'})}\cdot \prod_{k,a}\int_{\mathbb{R}} e^{-4\pi\{u_{ka}^2 - i u_{ka}(w_{ka}-\overline{w'_{ka}})\}}\,du_{ka},$$

where $W = (w_{ka})$, $W' = (w'_{ka}) \in \mathbb{C}^{(m,n)}$ and $U = (u_{ka}) \in \mathbb{R}^{(m,n)}$. It is easy to show that

$$\int_{\mathbb{R}} e^{-4\pi\{u_{ka}^2 - i u_{ka}(w_{ka}-\overline{w'_{ka}})\}}\,du_{ka} = e^{-\pi(w_{ka}-\overline{w'_{ka}})^2}.$$

Thus we get

$$\mathcal{I}(W,W') = e^{\pi\,\sigma(W\,{}^tW + \overline{W'}\,{}^t\overline{W'})}\cdot e^{-\pi\sum_{k,a}(w_{ka}-\overline{w'_{ka}})^2} \\ = e^{2\pi\sum_{k,a} w_{ka}\overline{w'_{ka}}} \\ = e^{2\pi\,\sigma(W\,{}^t\overline{W'})}.$$

□

For $U \in \mathbb{R}^{(m,n)}$ and $W \in \mathbb{C}^{(m,n)}$, we put

$$k_{\mathcal{M}}(U,W) := e^{2\pi\,\sigma\{\mathcal{M}(-U\,{}^tU-\frac{1}{2}W\,{}^tW+2U\,{}^tW)\}}. \tag{1.5.41}$$

Lemma 12. *Let $\mathcal{M}$ be a positive definite, symmetric half-integral matrix of degree m. Then we have*

$$k_{\mathcal{M}}(U,W) = k(\mathcal{M}^{\frac{1}{2}}U,\ -i\,\mathcal{M}^{\frac{1}{2}}W) \tag{1.5.42}$$

and

$$\int_{\mathbb{R}^{(m,n)}} k_{\mathcal{M}}(U,W)\,\overline{k_{\mathcal{M}}(U,W')}\,dU = (\det\mathcal{M})^{-\frac{n}{2}}\cdot e^{2\pi\sigma(\mathcal{M}W\,{}^t\overline{W'})}. \tag{1.5.43}$$

Proof. The formula (1.5.42) follows immediately from a straightforward computation. We put

$$\mathcal{I}_{\mathcal{M}}(W,W') := \int_{\mathbb{R}^{(m,n)}} k_{\mathcal{M}}(U,W)\,\overline{k_{\mathcal{M}}(U,W')}\,dU.$$

Using (1.5.42), we have

$$\begin{aligned}\mathcal{I}_{\mathcal{M}}(W,W') &= \int_{\mathbb{R}^{(m,n)}} k\left(\mathcal{M}^{\frac{1}{2}}U,\ -i\mathcal{M}^{\frac{1}{2}}W\right)\overline{k\left(\mathcal{M}^{\frac{1}{2}}U,-i\mathcal{M}^{\frac{1}{2}}W'\right)}\,dU\\ &= (\det\mathcal{M})^{-\frac{n}{2}}\int_{\mathbb{R}^{(m,n)}} k\left(U,-i\mathcal{M}^{\frac{1}{2}}W\right)\overline{k\left(U,-i\mathcal{M}^{\frac{1}{2}}W'\right)}\,dU\\ &= (\det\mathcal{M})^{-\frac{n}{2}}\cdot e^{2\pi\,\sigma(\mathcal{M}W\,{}^t\overline{W'})} \qquad \text{(by Lemma 11)}\end{aligned}$$

□

We recall that the Fock representation $U^{F,\mathcal{M}}$ of the real Heisenberg group G in $\mathcal{H}_{F,\mathcal{M}}$(cf. (1.5.23)) is given by

$$\left(U^{F,\mathcal{M}}(g)f\right)(W) = e^{2\pi i\,\sigma(\mathcal{M}\kappa)}\cdot e^{-\pi\,\sigma\{\mathcal{M}(\zeta\,{}^t\bar{\zeta}+2\,W\,{}^t\bar{\zeta})\}}\,f(W+\zeta), \tag{1.5.44}$$

where $g=(\lambda,\mu,\kappa)\in G,\ f\in\mathcal{H}_{F,\mathcal{M}}$ and $\zeta=\lambda+i\,\mu\in\mathbb{C}^{(m,n)}$.

Lemma 13. *The Fock representation $U^{F,\mathcal{M}}$ of G in $\mathcal{H}_{F,\mathcal{M}}$ is unitary.*

Proof. For brevity, we put $U_{g,f}(W) := \left(U^{F,\mathcal{M}}(g)f\right)(W)$ for $g=(\lambda,\mu,\kappa)\in G$ and $f\in\mathcal{H}_{F,\mathcal{M}}$. Then we have

$$
\begin{aligned}
&(U_{g,f},\, U_{g,f})_{\mathcal{M}} = \|U_{g,f}\|^2_{\mathcal{M}} \\
&= \int_{\mathbb{C}^{(m,n)}} U_{g,f}(W)\,\overline{U_{g,f}(W)}\, d\mu_{\mathcal{M}}(W) \\
&= \int_{\mathbb{C}^{(m,n)}} e^{-\pi\,\sigma\{\mathcal{M}(\zeta\,{}^t\bar{\zeta} + 2W\,{}^t\bar{\zeta} + \bar{\zeta}\,{}^t\zeta + 2\,\overline{W}\,{}^tW + 2\,W\,{}^t\overline{W})\}}\, |f(W+\zeta)|^2\, dW \\
&= \int_{\mathbb{C}^{(m,n)}} |f(W)|^2\, d\mu_{\mathcal{M}}(W) \\
&= (f, f)_{\mathcal{M}} = \|f\|^2_{\mathcal{M}}.
\end{aligned}
$$

□

We recall that the Schrödinger representation $U^{S,\mathcal{M}} := U_{\sigma_{\mathcal{M}}}$ of the real Heisenberg group G in the Hilbert space $\mathcal{H}_{S,\mathcal{M}} \cong L^2\left(\mathbb{R}^{(m,n)},\, d\xi\right)$ (cf. (1.4.8)) is given by

$$
\left(U^{S,\mathcal{M}}(g)f\right)(\xi) = e^{2\pi i\,\sigma\{\mathcal{M}(\kappa + \mu\,{}^t\lambda + 2\,\mu\,{}^t\xi)\}}\, f(\xi+\lambda), \tag{1.5.45}
$$

where $g = (\lambda, \mu, \kappa) \in G$, $f \in \mathcal{H}_{S,\mathcal{M}}$ and $\xi \in \mathbb{R}^{(m,n)}$. $U^{S,\mathcal{M}}$ is called the Schrödinger representation of G of index $\mathcal{M}$. The inner product $(\ ,\ )_{S,\mathcal{M}}$ on $\mathcal{H}_{S,\mathcal{M}}$ is given by

$$
(f_1,\, f_2)_{S,\mathcal{M}} = \int_{\mathbb{R}^{(m,n)}} f_1(U)\,\overline{f_2(U)}\, dU, \quad f_1, f_2 \in \mathcal{H}_{S,\mathcal{M}}.
$$

And we define the norm $\|\ \|_{S,\mathcal{M}}$ on $\mathcal{H}_{S,\mathcal{M}}$ by

$$
\|f\|^2_{S,\mathcal{M}} = \int_{\mathbb{R}^{(m,n)}} |f(U)|^2\, dU, \quad f \in \mathcal{H}_{S,\mathcal{M}}.
$$

Theorem 7. *The Fock representation $\left(U^{F,\mathcal{M}}, \mathcal{H}_{F,\mathcal{M}}\right)$ of G is untarily equivalent to the Schrödinger representation $\left(U^{S,\mathcal{M}},\, \mathcal{H}_{S,\mathcal{M}}\right)$ of G of index $\mathcal{M}$. Therefore the Fock representation $U^{F,\mathcal{M}}$ is irreducible. The intertwining unitary isometry $I_{\mathcal{M}} : \mathcal{H}_{S,\mathcal{M}} \longrightarrow \mathcal{H}^{F,\mathcal{M}}$ is given by*

$$
(\,I_{\mathcal{M}} f\,)(W) = \int_{\mathbb{R}^{(m,n)}} k_{\mathcal{M}}(\xi, W)\, f(\xi)\, d\xi, \tag{1.5.46}
$$

where $f \in \mathcal{H}_{S,\mathcal{M}} = L^2\left(\mathbb{R}^{(m,n)},\, d\xi\right)$, $W \in \mathbb{C}^{(m,n)}$ and $k_{\mathcal{M}}(\xi, W)$ is a function on $\mathbb{R}^{(m,n)} \times \mathbb{C}^{(m,n)}$ defined by (1.5.41).

Proof. For any $f \in \mathcal{H}_{S,\mathcal{M}} = L^2\left(\mathbb{R}^{(m,n)},\, d\xi\right)$, we define

$$
(I_{\mathcal{M}} f)(W) = \int_{\mathbb{R}^{(m,n)}} k_{\mathcal{M}}(\xi, W)\, f(\xi)\, d\xi, \quad W \in \mathbb{C}^{(m,n)}.
$$

Now we will show the following (I1), (I2) and (I3): (I1) The image of $\mathcal{H}_{S,\mathcal{M}}$ under $I_{\mathcal{M}}$ is contained in $\mathcal{H}_{F,\mathcal{M}}$.

(I2) $I_{\mathcal{M}}$ preserves the norms, i.e., $\|f\|_{S,\mathcal{M}} = \|I_{\mathcal{M}} f\|_{\mathcal{M}}$.

(I3) $I_{\mathcal{M}}$ is a bijective operator of $\mathcal{H}_{S,\mathcal{M}}$ onto $\mathcal{H}_{F,\mathcal{M}}$.
Before we prove (I1), (I2) and (I3), we prove the following lemma.

Lemma 14. *For a fixed* $U \in \mathbb{R}^{(m,n)}$, *we consider the Taylor expansion*

$$k_{\mathcal{M}}(U,W) = \sum_{J \in \mathbb{Z}_{\geq 0}^{(m,n)}} h_J(U)\, \Phi_{\mathcal{M},J}(W), \quad W \in \mathbb{C}^{(m,n)} \tag{1.5.47}$$

of the holomorphic function $k_{\mathcal{M}}(U,\,\cdot\,)$ *on* $\mathbb{C}^{(m,n)}$. *Then the set* $\left\{ h_J \,\middle|\, J \in \mathbb{Z}_{\geq 0}^{(m,n)} \right\}$ *forms a complete orthonormal system in* $L^2\left(\mathbb{R}^{(m,n)},\, d\xi\right)$. *Moreover, for a fixed* $W \in \mathbb{C}^{(m,n)}$, *(1.5.47) is the Fourier expansion of* $k_{\mathcal{M}}(\,\cdot\,,W)$ *with respect to this orthonormal system* $\left\{ h_J \,\middle|\, J \in \mathbb{Z}_{\geq 0}^{(m,n)} \right\}$.

Proof. Following Igusa [14], pp. 33-34, we can prove it. The detail will be left to the reader. □

If $f \in \mathcal{H}_{S,\mathcal{M}}$, then by the Schwarz inequality and Lemma 12, (1.5.43), we have

$$\begin{aligned} |\,(I_{\mathcal{M}} f\,)\,(W)| &\leq \left(\int_{\mathbb{R}^{(m,n)}} |k_{\mathcal{M}}(U,W)|^2\, dU \right)^{\frac{1}{2}} \cdot \left(\int_{\mathbb{R}^{(m,n)}} |f(U)|^2\, dU \right)^{\frac{1}{2}} \\ &= (\det \mathcal{M})^{-\frac{n}{4}} \cdot e^{\pi\, \sigma(\mathcal{M} W\, {}^t\overline{W})}\, \|f\|_{S,\mathcal{M}}. \end{aligned}$$

Thus the above integral $(I_{\mathcal{M}} f)(W)$ converges uniformly on any compact subset of $\mathbb{C}^{(m,n)}$ and hence $(I_{\mathcal{M}} f)(W)$ is holomorphic in $\mathbb{C}^{(m,n)}$. And according to Lemma 14, we get

$$\begin{aligned} (\,I_{\mathcal{M}} f\,)\,(W) &= \sum_{J \in \mathbb{Z}_{\geq 0}^{(m,n)}} \int_{\mathbb{R}^{(m,n)}} h_J(U)\, f(U)\, \Phi_{\mathcal{M},J}(W)\, dU \\ &= \sum_{J \in \mathbb{Z}_{\geq 0}^{(m,n)}} (h_J,\, \bar{f}\,)_{S,\mathcal{M}}\, \Phi_{\mathcal{M},J}(W). \end{aligned}$$

Therefore we get

$$\begin{aligned}
|I_{\mathcal{M}} f\|_{F,\mathcal{M}}^2 &= \int_{\mathbb{C}^{(m,n)}} |I_{\mathcal{M}} f(W)|^2 \, d\mu_{\mathcal{M}}(W) \\
&= \sum_{J,\,K\in\mathbb{Z}_{\geq 0}^{(m,n)}} (h_J,\, \bar{f}\,)_{S,\mathcal{M}} \cdot \overline{(h_K,\, \bar{f}\,)} \int_{\mathbb{C}^{(m,n)}} \Phi_{\mathcal{M},J}(W)\, \overline{\Phi_{\mathcal{M},K}(W)}\, d\mu_{\mathcal{M}}(W) \\
&= \sum_{J\in\mathbb{Z}_{\geq 0}^{(m,n)}} |(h_j,\, \bar{f}\,)_{S,\mathcal{M}}|^2 \qquad (\text{by } (6.37)) \\
&= \|f\|_{S,\mathcal{M}}^2 < \infty.
\end{aligned}$$

This proves (I1) and (I2). It is easy to see that $I_{\mathcal{M}}\overline{h_J} = \Phi_{\mathcal{M},J}$ for all $J \in \mathbb{Z}_{\geq 0}^{(m,n)}$. Since the set $\left\{ \Phi_{\mathcal{M},J} \,\middle|\, J \in \mathbb{Z}_{\geq 0}^{(m,n)} \right\}$ forms a complete orthonormal system of $\mathcal{H}_{F,\mathcal{M}}$, $I_{\mathcal{M}}$ is surjective. Obviously the injectivity of $I_{\mathcal{M}}$ follows immediately from the fact that $I_{\mathcal{M}}\overline{h_J} = \Phi_{\mathcal{M},J}$ for all $J \in \mathbb{Z}_{\geq 0}^{(m,n)}$. This proves (I3).

On the other hand, we let $f \in \mathcal{H}_{S,\mathcal{M}}$ and $g = (\lambda, \mu, \kappa) \in G$. We put $\zeta = \lambda + i\,\mu$. Then we get

$$\begin{aligned}
&\left(U^{F,\mathcal{M}}(g)(I_{\mathcal{M}} f)\right)(W) \\
&= e^{2\pi i\,\sigma(\mathcal{M}\kappa)} \cdot e^{-\pi\,\sigma\{\mathcal{M}(\zeta\,{}^t\bar{\zeta} + 2\,W\,{}^t\bar{\zeta})\}}\, (I_{\mathcal{M}} f)(W+\zeta) \quad (\,\text{by } (6.44)\,) \\
&= e^{2\pi i\,\sigma(\mathcal{M}\kappa)} \cdot e^{-\pi\sigma\{\mathcal{M}(\zeta\,{}^t\bar{\zeta} + 2\,W\,{}^t\bar{\zeta})\}} \int_{\mathbb{R}^{(m,n)}} k_{\mathcal{M}}(U, W+\zeta)\, f(U)\, dU.
\end{aligned}$$

We define the function $A_{\mathcal{M}} : \mathbb{R}^{(m,n)} \times \mathbb{R}^{(m,n)} \longrightarrow \mathbb{C}$ by

$$A_{\mathcal{M}}(U,W) := \sigma\left\{ \mathcal{M}\left(-U\,{}^tU - \frac{W\,{}^tW}{2} + 2U\,{}^tW \right) \right\}. \tag{1.5.48}$$

Obviously $\kappa_{\mathcal{M}}(U,W) = e^{2\pi A_{\mathcal{M}}(U,W)}$ for $U \in \mathbb{R}^{(m,n)}$ and $W \in \mathbb{C}^{(m,n)}$.

By an easy computation, we get

$$A_{\mathcal{M}}(U, W+\zeta) - A_{\mathcal{M}}(U-\lambda, W) = \sigma\left\{ \mathcal{M}\left(\frac{\zeta\,{}^t\bar{\zeta}}{2} + W\,{}^t\bar{\zeta} - i\,\lambda\,{}^t\mu + 2\,i\,U\,{}^t\mu \right) \right\}.$$

Therefore we get

$$\begin{aligned}
&k_{\mathcal{M}}(U, W+\zeta) \\
&= e^{2\pi A_{\mathcal{M}}(U-\lambda, W)} \cdot e^{2\pi\,\sigma\left\{\mathcal{M}\left(\frac{1}{2}\,\zeta\,{}^t\bar{\zeta} + W\,{}^t\bar{\zeta} - i\,\lambda\,{}^t\mu + 2\,i\,U\,{}^t\mu\right)\right\}} \\
&= k_{\mathcal{M}}(U-\lambda, W) \cdot e^{2\pi\,\sigma\left\{\mathcal{M}\left(\frac{1}{2}\,\zeta\,{}^t\bar{\zeta} + W\,{}^t\bar{\zeta} - i\,\lambda\,{}^t\mu + 2\,i\,U\,{}^t\mu\right)\right\}}.
\end{aligned}$$

Hence we have

$$
\begin{aligned}
&\left(U^{F,\mathcal{M}}(g)(I_{\mathcal{M}}f)\right)(W) \\
&= \int_{\mathbb{R}^{(m,n)}} e^{2\pi i\,\sigma\left\{\mathcal{M}(\kappa+2\,U\,{}^t\mu-\lambda\,{}^t\mu)\right\}}\, k_{\mathcal{M}}(U-\lambda,W)\, f(U)\, dU \\
&= \int_{\mathbb{R}^{(m,n)}} e^{2\pi i\,\sigma\left\{\mathcal{M}(\kappa+2\,\lambda\,{}^t\mu+2\,U\,{}^t\mu-\lambda\,{}^t\mu)\right\}}\, k_{\mathcal{M}}(U,W)\, f(U+\lambda)\, dU \\
&= \int_{\mathbb{R}^{(m,n)}} e^{2\pi i\,\sigma\left\{\mathcal{M}(\kappa+2\,U\,{}^t\mu+\lambda\,{}^t\mu)\right\}}\, k_{\mathcal{M}}(U,W)\, f(U+\lambda)\, dU \\
&= \int_{\mathbb{R}^{(m,n)}} k_{\mathcal{M}}(U,W)\,\left(U^{S,\mathcal{M}}(g)f\right)(U)\, dU \qquad \text{(by (6.45))} \\
&= \left(I_{\mathcal{M}}\left(U^{S,\mathcal{M}}(g)f\right)\right)(W).
\end{aligned}
$$

So far we proved that $U^{F,\mathcal{M}} \circ I_{\mathcal{M}} = I_{\mathcal{M}} \circ U^{S,\mathcal{M}}(g)$ for all $g \in G$. That is, the unitary isometry $I_{\mathcal{M}}$ of $\mathcal{H}_{S,\mathcal{M}}$ onto $\mathcal{H}_{F,\mathcal{M}}$ is the intertwining operator. This completes the proof.

□

The infinitesimal representation $dU^{F,\mathcal{M}}$ associated to the Fock representation $U^{F,\mathcal{M}}$ is given as follows.

Proposition 5. *Let $\mathcal{M}$ be as before. We put*

$$\mathcal{M} = (\,\mathcal{M}_{kl}\,), \quad (2\pi\mathcal{M})^{\frac{1}{2}} = (\tau_{kl}),$$

where $\tau_{kl} \in \mathbb{R}$ and $1 \le k,l \le m$. For each $J = (J_{ka}) \in \mathbb{Z}^{(m,n)}_{\ge 0}$ and $W = (W_{ka}) \in \mathbb{C}^{(m,n)}$, we have

$$dU^{F,\mathcal{M}}(D^0_{kl})\,\Phi_{\mathcal{M},J}(W) = 2\,\pi\, i\,\mathcal{M}_{kl}\,\Phi_{\mathcal{M},J}(W), \quad 1 \le k \le l \le m. \tag{1.5.49}$$

$$
\begin{aligned}
dU^{F,\mathcal{M}}(D_{ka})\,\Phi_{\mathcal{M},J}(W) &= -2\,\pi\left(\sum_{p=1}^m \mathcal{M}_{pk}W_{pa}\right)\Phi_{\mathcal{M},J}(W) \\
&\quad + \sum_{p=1}^m \tau_{pk}\, J_{pa}^{\frac{1}{2}}\,\Phi_{\mathcal{M},J-\epsilon_{pa}}(W).
\end{aligned} \tag{1.5.50}
$$

$$
\begin{aligned}
dU^{F,\mathcal{M}}(\widehat{D}_{lb}\,\Phi_{\mathcal{M},J}(W) &= 2\,\pi\, i\left(\sum_{p=1}^m \mathcal{M}_{pl}W_{pb}\right)\Phi_{\mathcal{M},J}(W) \\
&\quad + i\sum_{p=1}^m \tau_{pl}\, J_{pb}^{\frac{1}{2}}\,\Phi_{\mathcal{M},J-\epsilon_{lb}}(W).
\end{aligned} \tag{1.5.51}
$$

Proof. We put $E^0_{kl} = \frac{1}{2}(E_{kl} + E_{lk})$, where $1 \le k \le l \le m$.

$$\begin{aligned} dU^{F,\mathcal{M}}(D^0_{kl})\,\Phi_{\mathcal{M},J}(W) &= \frac{d}{dt}\bigg|_{t=0} U^{F,\mathcal{M}}(\exp tX^0_{kl})\,\Phi_{\mathcal{M},J}(W) \\ &= \frac{d}{dt}\bigg|_{t=0} U^{F,\mathcal{M}}\left((0,0,tE^0_{kl})\right)\,\Phi_{\mathcal{M},J}(W) \\ &= \lim_{t\to 0}\frac{e^{2\pi i\,\sigma(t\mathcal{M}E^0_{kl})}-I}{t}\,\Phi_{\mathcal{M},J}(W) \\ &= \lim_{t\to 0}\frac{e^{2\pi i t\mathcal{M}_{kl}}-I}{t}\,\Phi_{\mathcal{M},J}(W) \\ &= 2\,\pi\, i\,\mathcal{M}_{kl}\,\Phi_{\mathcal{M},J}(W). \end{aligned}$$

And we have

$$\begin{aligned} & dU^{F,\mathcal{M}}(D_{ka})\,\Phi_{\mathcal{M},J}(W) \\ &= \frac{d}{dt}\bigg|_{t=0} U^{F,\mathcal{M}}(\exp tX_{ka})\,\Phi_{\mathcal{M},J}(W) \\ &= \frac{d}{dt}\bigg|_{t=0} U^{F,\mathcal{M}}\left((tE_{ka},0,0)\right)\Phi_{\mathcal{M},J}(W) \\ &= \frac{d}{dt}\bigg|_{t=0} e^{-\pi t^2\,\sigma(\mathcal{M}E_{ka}\,{}^tE_{ka})-2\pi t\,\sigma(\mathcal{M}W\,{}^tE_{ka})}\,\Phi_{\mathcal{M},J}(W+tE_{ka}) \\ &= -2\,\pi\left(\sum_{p=1}^m \mathcal{M}_{pk}W_{pa}\right)\Phi_{\mathcal{M},J}(W) + \frac{d}{dt}\bigg|_{t=0}\Phi_{\mathcal{M},J}(W+tE_{ka}) \\ &= -2\,\pi\left(\sum_{p=1}^m \mathcal{M}_{pk}W_{pa}\right)\Phi_{\mathcal{M},J}(W) + \sum_{p=1}^m \tau_{pk}\,J_{pa}^{\frac12}\,\Phi_{\mathcal{M},J-\epsilon_{pa}}(W). \end{aligned}$$

Finally we obtain

$$\begin{aligned} & dU^{F,\mathcal{M}}(\widehat{D}_{lb})\,\Phi_{\mathcal{M},J}(W) \\ &= \frac{d}{dt}\bigg|_{t=0} U^{F,\mathcal{M}}(\exp t\widehat{X}_{lb})\Phi_{\mathcal{M},J}(W) \\ &= \frac{d}{dt}\bigg|_{t=0} U^{F,\mathcal{M}}\left((0,tE_{lb},0)\right)\Phi_{\mathcal{M},J}(W) \\ &= \frac{d}{dt}\bigg|_{t=0} e^{-\pi t^2\,\sigma(\mathcal{M}E_{lb}\,{}^tE_{lb})\,+\,2\pi i t\,\sigma(\mathcal{M}W\,{}^tE_{lb})}\,\Phi_{\mathcal{M},J}(W+i\,tE_{lb}) \\ &= 2\pi i\left(\sum_{p=1}^m \mathcal{M}_{pl}W_{pb}\right)\Phi_{\mathcal{M},J}(W) + \frac{d}{dt}\bigg|_{t=0}\Phi_{\mathcal{M},J}(W+i\,tE_{lb}) \\ &= 2\pi i\left(\sum_{p=1}^m \mathcal{M}_{pl}W_{pb}\right)\Phi_{\mathcal{M},J}(W) + i\sum_{p=1}^m \tau_{pl}\,J_{pb}^{\frac12}\,\Phi_{\mathcal{M},J-\epsilon_{pb}}(W). \end{aligned}$$

□

1.6 Lattice Representations

Let $L := \mathbb{Z}^{(m,n)} \times \mathbb{Z}^{(m,n)}$ be the lattice in the vector space $V \cong \mathbb{C}^{(m,n)}$. Let B be an alternating bilinear form on V such that $B(L,L) \subset \mathbb{Z}$, that is, $\mathbb{Z}$-valued on $L \times L$. The dual L_B^* of L with respect to B is defined by

$$L_B^* := \{ v \in V \mid B(v,L) \in \mathbb{Z} \ for\ all\ l \in L \}.$$

Then $L \subset L_B^*$. If B is nondegenerate, L_B^* is also a lattice in V, called the *dual lattice* of L. In case B is nondegenerate, there exist a $\mathbb{Z}$-basis $\{ \xi_{11}, \xi_{12}, \cdots, \xi_{mn}, \eta_{11}, \eta_{12}, \cdots, \eta_{mn} \}$ of L and a set $\{ e_{11}, e_{12}, \cdots, e_{mn} \}$ of positive integers with $e_{11}|e_{12}$, $e_{12}|e_{13}, \cdots, e_{m,n-1}|e_{mn}$ such that

$$\begin{pmatrix} B(\xi_{ka}, \xi_{lb}) & B(\xi_{ka}, \eta_{lb}) \\ B(\eta_{ka}, \xi_{lb}) & B(\eta_{ka}, \eta_{lb} \end{pmatrix} = \begin{pmatrix} 0 & e \\ -e & 0 \end{pmatrix},$$

where $1 \leq k, l \leq m$, $1 \leq a, b \leq n$ and $e := \mathrm{diag}\,(e_{11}, e_{12}, \cdots, e_{mn})$ is the diagonal matrix of degree mn with entries $e_{11}, e_{12}, \cdots, e_{mn}$. It is well known that $[L_B^* : L] = (\det e)^2 = (e_{11}e_{12} \cdots e_{mn})^2$ (cf. [14] p. 72). The number $\det e$ is called the Pfaffian of B.

Now we consider the following subgroups of G:

$$\Gamma_L = \left\{ (\lambda, \mu, \kappa) \in G \mid (\lambda, \mu) \in L,\ \kappa \in \mathbb{R}^{(m,m)} \right\} \tag{1.6.1}$$

and

$$\Gamma_{L_B^*} = \left\{ (\lambda, \mu, \kappa) \in G \mid (\lambda, \mu) \in L_B^*,\ \kappa \in \mathbb{R}^{(m,m)} \right\}. \tag{1.6.2}$$

Then both Γ_L and $\Gamma_{L_B^*}$ are the normal subgroups of G.

We put

$$\mathcal{Z}_0 = \left\{ (0,0,\kappa) \in \mathcal{Z} \mid \kappa = {}^t\kappa \in \mathbb{Z}^{(m,m)} \text{ integral} \right\}. \tag{1.6.3}$$

It is easy to show that

$$\Gamma_{L_B^*} = \{ g \in G \mid g\gamma g^{-1}\gamma^{-1} \in \mathcal{Z}_0 \text{ for all } \gamma \in \Gamma_L \}.$$

We define

$$Y_L = \{ \phi \in \mathrm{Hom}\,(\Gamma_L, \mathbb{C}_1^\times) \mid \phi \text{ is trivial on } \mathcal{Z}_0 \} \tag{1.6.4}$$

and

$$Y_{L,S} = \left\{ \phi \in Y_L \mid \phi(\kappa) = e^{2\pi i \sigma(S\kappa)} \text{ for all } \kappa = {}^t\kappa \in \mathbb{R}^{(m,m)} \right\} \tag{1.6.5}$$

for each symmetric real matrix S of degree m. We observe that if S is not half-integral, then $Y_L = \emptyset$ and so $Y_{L,S} = \emptyset$. It is clear that if S is symmetric half-integral, then $Y_{L,S}$ is not empty. Thus we have

$$Y_L = \cup_{\mathcal{M}} Y_{L,\mathcal{M}}, \tag{1.6.6}$$

where $\mathcal{M}$ runs through the set of all symmetric half-integral matrices of degree m.

Lemma 15. *Let $\mathcal{M}$ be a symmetric half-integral matrix of degree m with $\mathcal{M} \neq 0$. Then any element ϕ of $Y_{L,\mathcal{M}}$ is of the form $\phi_{\mathcal{M},q}$. Here $\phi_{\mathcal{M},q}$ is the character of Γ_L defined by*

$$\phi_{\mathcal{M},q}((l,\kappa)) := e^{2\pi i\,\sigma(\mathcal{M}\kappa)} \cdot e^{\pi i\,q(l)}, \quad (l,\kappa) \in \Gamma_L, \tag{1.6.7}$$

where $q : L \longrightarrow \mathbb{R}/2\mathbb{Z} \cong [0,2)$ is a function on L satisfying the following condition:

$$q(l_0 + l_1) \equiv q(l_0) + q(l_1) - 2\sigma\{\mathcal{M}(\lambda_0\,{}^t\mu_1 - \mu_0\,{}^t\lambda_1)\} \mod 2 \tag{1.6.8}$$

for all $l_0 = (\lambda_0, \mu_0) \in L$ and $l_1 = (\lambda_1, \mu_1) \in L$.

Proof. (1.6.8) follows immediately from the fact that $\phi_{\mathcal{M},q}$ is a character of Γ_L. It is obvious that any element of $Y_{L,\mathcal{M}}$ is of the form $\phi_{\mathcal{M},q}$. □

Lemma 16. *An element of $Y_{L,0}$ is of the form $\phi_{k,l}$ $(k, l \in \mathbb{R}^{(m,n)})$. Here $\phi_{k,l}$ is the character of Γ_L defined by*

$$\phi_{k,l}(\gamma) := e^{2\pi i\,\sigma(k\,{}^t\lambda + l\,{}^t\mu)}, \quad \gamma = (\lambda, \mu, \kappa) \in \Gamma_L. \tag{1.6.9}$$

Proof. It is easy to prove it and so we omit the proof. □

Lemma 17. *Let $\mathcal{M}$ be a nonsingular symmetric half-integral matrix of degree m. Let $\phi_{\mathcal{M},q_1}$ and $\phi_{\mathcal{M},q_2}$ be the characters of Γ_L defined by (1.6.7). The character ϕ of Γ_L defined by $\phi := \phi_{\mathcal{M},q_1} \cdot \phi_{\mathcal{M},q_2}^{-1}$ is an element of $Y_{L,0}$.*

Proof. It follows from the fact that there exists an element $g = (\mathcal{M}^{-1}\lambda, \mathcal{M}^{-1}\mu, 0) \in G$ with $(\lambda, \mu) \in V$ such that

$$\phi_{\mathcal{M},q_1}(\gamma) = \phi_{\mathcal{M},q_2}(g\gamma g^{-1}) \quad \text{for all } \gamma \in \Gamma_L.$$

□

We note that the alternating bilinear form $\mathbf{A}$ on V defined by (1.5.2) is nondegenerate and $\mathbb{Z}$-valued on $L \times L$. According to (1.5.3), the elementary divisors $e_{11}, e_{12}, \cdots, e_{mn}$ of $\mathbf{A}$ are all one and L is self-dual, i.e., $L = L^*_{\mathbf{A}}$. The set

$$\{\, P_{11}, P_{12}, \cdots, P_{mn}, Q_{11}, Q_{12}, \cdots, Q_{mn} \,\}$$

forms a symplectic basis of V with respect to $\mathbf{A}$. We fix a coordinate $P_{11}, \cdots, P_{mn}, Q_{11}, \cdots, Q_{mn}$ on V.

For a unitary character $\varphi_{\mathcal{M},q}$ of Γ_L defined by (1.6.7), we let

$$\pi_{\mathcal{M},q} = \operatorname{Ind}_{\Gamma_L}^{G} \varphi_{\mathcal{M},q} \tag{1.6.10}$$

be the representation of G induced from $\varphi_{\mathcal{M},q}$. Let $\mathcal{H}_{\mathcal{M},q}$ be the Hilbert space consisting of all measurable functions $\phi : G \longrightarrow \mathbb{C}$ satisfying

(L1) $\phi(\gamma g) = \varphi_{\mathcal{M},q}(\gamma)\, \phi(g)$ for all $\gamma \in \Gamma_L$ and $g \in G$.

(L2) $\|\phi\|^2_{\mathcal{M},q} = \int_{\Gamma_L\backslash G} |\phi(\bar{g})|\, d\bar{g} l \infty, \quad \bar{g} = \Gamma_L g.$

The induced representation $\pi_{\mathcal{M},q}$ is realized in $\mathcal{H}_{\mathcal{M},q}$ as follows:

$$\left(\pi_{\mathcal{M},q}(g_0)\phi \right)(g) = \phi(gg_0), \quad g_0, g \in G, \ \phi \in \mathcal{H}_{\mathcal{M},q}. \tag{1.6.11}$$

$\pi_{\mathcal{M},q}$ is called the **lattice representation** of G associated with the lattice L.

Theorem 8. *Let $\mathcal{M}$ be a positive definite, symmetric half integral matrix of degree m. Let $\varphi_{\mathcal{M}}$ be the character of Γ_L defined by $\varphi_{\mathcal{M}}((\lambda, \mu, \kappa)) := e^{2\pi i\, \sigma(\mathcal{M}\kappa)}$ for all $(\lambda, \mu, \kappa) \in \Gamma_L$. Then the representation*

$$\pi_{\mathcal{M}} := \mathrm{Ind}_{\Gamma_L}^G \varphi_{\mathcal{M}} \tag{1.6.12}$$

induced from the character $\varphi_{\mathcal{M}}$ is unitarily equivalent to the representation

$$\bigoplus U_{\mathcal{M}} := \bigoplus \mathrm{Ind}_K^G \sigma_{\mathcal{M}} \quad ((\det 2\mathcal{M})^n\text{-copies}),$$

where K (resp. $\sigma_{\mathcal{M}}$) is defined by (1.4.1) (resp. (1.4.6)).

Proof. We first recall that the induced representation $\pi_{\mathcal{M}}$ is realized in the Hilbert space $\mathcal{H}_{\mathcal{M}}$ consisting of all measurable functions $\phi : G \longrightarrow \mathbb{C}$ satisfying the conditions

$$\phi((\lambda_0, \mu_0, \kappa_0) \circ g) = e^{2\pi i\, \sigma(\mathcal{M}\kappa_0)}\, \phi(g), \quad (\lambda_0, \mu_0, \kappa_0) \in \Gamma_L, \ g \in G \tag{1.6.13}$$

and

$$\|\phi\|^2_{\pi,\mathcal{M}} := \int_{\Gamma_L\backslash G} |\phi(\bar{g})|^2\, d\bar{g} l \infty, \quad \bar{g} = \Gamma_L \circ g. \tag{1.6.14}$$

Now we write

$$g_0 = [\lambda_0, \mu_0, \kappa_0] \in \Gamma_L \ \text{ and } \ g = [\lambda, \mu, \kappa] \in G.$$

For $\phi \in \mathcal{H}_{\mathcal{M}}$, we have

$$\phi(g_0 \diamond g) = \phi([\lambda_0 + \lambda, \mu_0 + \mu, \kappa_0 + \kappa + \lambda_0\, {}^t\mu + \mu\, {}^t\lambda_0]). \tag{1.6.15}$$

On the other hand, we get

$$\begin{aligned} \phi(g_0 \diamond g) &= \phi((\lambda_0, \mu_0, \kappa_0 - \mu_0\, {}^t\lambda_0) \circ g) \\ &= e^{2\pi i\, \sigma\{\mathcal{M}(\kappa_0 - \mu_0\, {}^t\lambda_0)\}}\, \phi(g) \\ &= e^{2\pi i\, \sigma(\mathcal{M}\kappa_0)}\, \phi(g) \quad (\text{because } \sigma(\mathcal{M}\mu_0\, {}^t\lambda_0) \in \mathbb{Z}) \end{aligned}$$

Thus putting $\kappa' := \kappa_0 + \lambda_0\, {}^t\mu + \mu\, {}^t\lambda_0$, we get

$$\phi([\lambda_0 + \lambda, \mu_0 + \mu, \kappa + \kappa']) = e^{2\pi i\, \sigma(\mathcal{M}\kappa')} \cdot e^{-4\pi i\, \sigma(\mathcal{M}\lambda_0\, {}^t\mu)}\, \phi([\lambda, \mu, \kappa]). \quad (1.6.16)$$

Putting $\lambda_0 = \kappa_0 = 0$ in (1.6.16), we have

$$\phi([\lambda, \mu + \mu_0, \kappa]) = \phi([\lambda, \mu, \kappa]) \ \text{ for all } \mu_0 \in \mathbb{Z}^{(m,n)} \text{ and } [\lambda, \mu, \kappa] \in G. \quad (1.6.17)$$

Therefore if we fix λ and κ, ϕ is periodic in μ with respect to the lattice $\mathbb{Z}^{(m,n)}$ in $\mathbb{R}^{(m,n)}$. We note that

$$\phi([\lambda, \mu, \kappa]) = \phi([0, 0, \kappa] \diamond [\lambda, \mu, 0]) = e^{2\pi i\, \sigma(\mathcal{M}\kappa)}\, \phi([\lambda, \mu, 0])$$

for $[\lambda, \mu, \kappa] \in G$. Hence ϕ admits a Fourier expansion in μ :

$$\phi([\lambda, \mu, \kappa]) = e^{2\pi i\, \sigma(\mathcal{M}\kappa)} \sum_{N \in \mathbb{Z}^{(m,n)}} c_N(\lambda)\, e^{2\pi i\, \sigma(N\, {}^t\mu)}. \quad (1.6.18)$$

If $\lambda_0 \in \mathbb{Z}^{(m,n)}$, then we have

$$\begin{aligned}
\phi([\lambda + \lambda_0, \mu, \kappa]) &= e^{2\pi i\, \sigma(\mathcal{M}\kappa)} \sum_{N \in \mathbb{Z}^{(m,n)}} c_N(\lambda + \lambda_0)\, e^{2\pi i\, \sigma(N\, {}^t\mu)} \\
&= e^{-4\pi i\, \sigma(\mathcal{M}\lambda_0\, {}^t\mu)}\, \phi([\lambda, \mu, \kappa]) \qquad (\text{by } (1.6.16)) \\
&= e^{-4\pi i\, \sigma(\mathcal{M}\lambda_0\, {}^t\mu)}\, e^{2\pi i\, \sigma(\mathcal{M}\kappa)} \sum_{N \in \mathbb{Z}^{(m,n)}} c_N(\lambda)\, e^{2\pi i\, \sigma(N\, {}^t\mu)}, \\
&= e^{2\pi i\, \sigma(\mathcal{M}\kappa)} \sum_{N \in \mathbb{Z}^{(m,n)}} c_N(\lambda)\, e^{2\pi i\, \sigma\{(N - 2\mathcal{M}\lambda_0)\, {}^t\mu\}} \ (\text{by } (1.6.18))
\end{aligned}$$

So we get

$$\begin{aligned}
&\sum_{N \in \mathbb{Z}^{(m,n)}} c_N(\lambda + \lambda_0)\, e^{2\pi i\, \sigma(N\, {}^t\mu)} \\
&= \sum_{N \in \mathbb{Z}^{(m,n)}} c_N(\lambda)\, e^{2\pi i\, \sigma\{(N - 2\mathcal{M}\lambda_0)\, {}^t\mu\}} \\
&= \sum_{N \in \mathbb{Z}^{(m,n)}} c_{N + 2\mathcal{M}\lambda_0}(\lambda)\, e^{2\pi i\, \sigma(N\, {}^t\mu)}.
\end{aligned}$$

Hence we get

$$c_N(\lambda + \lambda_0) = c_{N + 2\mathcal{M}\lambda_0}(\lambda) \ \text{ for all } \lambda_0 \in \mathbb{Z}^{(m,n)} \text{ and } \lambda \in \mathbb{R}^{(m,n)}. \quad (1.6.19)$$

Consequently, it is enough to know only the coefficients $c_\alpha(\lambda)$ for the representatives α in $\mathbb{Z}^{(m,n)}$ modulo $2\mathcal{M}$. It is obvious that the number of all such representatives α's is $(\det 2\mathcal{M})^n$. We denote by $\mathcal{J}$ a complete system of such representatives α's in $\mathbb{Z}^{(m,n)}$ modulo $2\mathcal{M}$. Then we have

$$
\begin{aligned}
&\phi([\lambda,\mu,\kappa])\\
&= e^{2\pi i\,\sigma(\mathcal{M}\kappa)}\Big\{\sum_{N\in\mathbb{Z}^{(m,n)}} c_{\alpha+2\mathcal{M}N}(\lambda)\,e^{2\pi i\,\sigma\{(\alpha+2\mathcal{M}N)\,{}^t\mu\}}\\
&\qquad + \sum_{N\in\mathbb{Z}^{(m,n)}} c_{\beta+2\mathcal{M}N}(\lambda)\,e^{2\pi i\sigma\{(\beta+2\mathcal{M}N)\,{}^t\mu\}}\\
&\qquad \vdots\\
&\qquad + \sum_{N\in\mathbb{Z}^{(m,n)}} c_{\gamma+2\mathcal{M}N}(\lambda)\,e^{2\pi i\,\sigma\{(\gamma+2\mathcal{M}N)\,{}^t\mu\}}\Big\},
\end{aligned}
$$

where $\{\alpha,\beta,\cdots,\gamma\}$ denotes the complete system $\mathcal{J}$.

For each $\alpha\in\mathcal{J}$, we denote by $\mathcal{H}_{\mathcal{M},\alpha}$ the Hilbert space consisting of Fourier expansions

$$
e^{2\pi i\,\sigma(\mathcal{M}\kappa)}\sum_{N\in\mathbb{Z}^{(m,n)}} c_{\alpha+2\mathcal{M}N}(\lambda)\,e^{2\pi i\,\sigma\{\,(\alpha+2\mathcal{M}N)\,{}^t\mu\}},\quad (\lambda,\mu,\kappa)\in G,
$$

where $c_N(\lambda)$ denotes the coefficients of the Fourier expansion (1.6.18) of $\phi\in\mathcal{H}_{\mathcal{M}}$ and ϕ runs over the set $\{\,\phi\in\pi_{\mathcal{M}}\,\}$. It is easy to see that $\mathcal{H}_{\mathcal{M},\alpha}$ is invariant under $\pi_{\mathcal{M}}$. We denote the restriction of $\pi_{\mathcal{M}}$ to $\mathcal{H}_{\mathcal{M},\alpha}$ by $\pi_{\mathcal{M},\alpha}$. Then we have

$$
\pi_{\mathcal{M}}=\bigoplus_{\alpha\in\mathcal{J}}\pi_{\mathcal{M},\alpha}. \tag{1.6.20}
$$

Let $\phi_\alpha\in\pi_{\mathcal{M},\alpha}$. Then for $[\lambda,\mu,\kappa]\in G$, we get

$$
\phi_\alpha([\lambda,\mu,\kappa])=e^{2\pi i\,\sigma(\mathcal{M}\kappa)}\sum_{N\in\mathbb{Z}^{(m,n)}} c_{\alpha+2\mathcal{M}N}(\lambda)\,e^{2\pi i\,\sigma\{(\alpha+2\mathcal{M}N)\,{}^t\mu\}}. \tag{1.6.21}
$$

We put

$$
I_\lambda=\overbrace{[0,1]\times[0,1]\times\cdots\times[0,1]}^{(m\times n)\text{-times}}\subset\Big\{\,[\lambda,0,0]\mid\lambda\in\mathbb{R}^{(m,n)}\Big\}
$$

and

$$
I_\mu=\overbrace{[0,1]\times[0,1]\times\cdots\times[0,1]}^{(m\times n)\text{-times}}\subset\Big\{\,[0,\mu,0]\mid\mu\in\mathbb{R}^{(m,n)}\Big\}.
$$

Then we obtain

$$
\int_{I_\mu}\phi_\alpha([\lambda,\mu,\kappa])\,e^{-2\pi i\,\sigma(\alpha\,{}^t\mu)}\,d\mu=e^{2\pi i\,\sigma(\mathcal{M}\kappa)}\,c_\alpha(\lambda),\quad \alpha\in\mathcal{J}. \tag{1.6.22}
$$

Since $\Gamma_L\backslash G\cong I_\lambda\times I_\mu$, we get

$$\begin{aligned}
\|\phi_\alpha\|^2_{\pi,\mathcal{M},\alpha} := \|\phi_\alpha\|^2_{\pi,\mathcal{M}} &= \int_{\Gamma_L\backslash G} |\phi_\alpha(\bar g)|^2\, d\bar g \\
&= \int_{I_\lambda}\int_{I_\mu} |\phi_\alpha(\bar g)|^2\, d\lambda d\mu \\
&= \int_{I_\lambda\times I_\mu} \left| \sum_{N\in\mathbb{Z}^{(m,n)}} c_{\alpha+2\mathcal{M}N}(\lambda)\, e^{2\pi i\,\sigma\{(\alpha+2\mathcal{M}N)\,{}^t\mu\}} \right|^2 d\lambda d\mu \\
&= \int_{I_\lambda} \sum_{N\in\mathbb{Z}^{(m,n)}} |c_{\alpha+2\mathcal{M}N}(\lambda)|^2\, d\lambda \\
&= \int_{I_\lambda} \sum_{N\in\mathbb{Z}^{(m,n)}} |c_\alpha(\lambda+N)|^2\, d\lambda \qquad (\mathit{by}\ (1.6.19)) \\
&= \int_{\mathbb{R}^{(m,n)}} |c_\alpha(\lambda)|^2\, d\lambda.
\end{aligned}$$

Since $\phi_\alpha \in \pi_{\mathcal{M},\alpha}$, $\|\phi_\alpha\|_{\pi,\mathcal{M},\alpha} < \infty$ and so $c_\alpha(\lambda) \in L^2\left(\mathbb{R}^{(m,n)}, d\xi\right)$ for all $\alpha \in \mathcal{J}$.

For each $\alpha \in \mathcal{J}$, we define the mapping $\vartheta_{\mathcal{M},\alpha}$ on $L^2(\mathbb{R}^{(m,n)}, d\xi)$ by

$$\begin{aligned}
(\vartheta_{\mathcal{M},\alpha} f)([\lambda,\mu,\kappa]) = e^{2\pi i\,\sigma(\mathcal{M}\kappa)} \\
\times \sum_{N\in\mathbb{Z}^{(m,n)}} f(\lambda+N)\, e^{2\pi i\,\sigma\{(\alpha+2\mathcal{M}N)\,{}^t\mu\}},
\end{aligned} \tag{1.6.23}$$

where $f \in L^2\left(\mathbb{R}^{(m,n)}, d\xi\right)$ and $[\lambda,\mu,\kappa] \in G$.

Lemma 18. *For each $\alpha \in \mathcal{J}$, the image of $L^2\left(\mathbb{R}^{(m,n)}, d\xi\right)$ under $\vartheta_{\mathcal{M},\alpha}$ is contained in $\mathcal{H}_{\mathcal{M},\alpha}$. Moreover, the mapping $\vartheta_{\mathcal{M},\alpha}$ is a one-to-one unitary operator of $L^2\left(\mathbb{R}^{(m,n)}, d\xi\right)$ onto $\mathcal{H}_{\mathcal{M},\alpha}$ preserving the norms. In other words, the mapping*

$$\vartheta_{\mathcal{M},\alpha} : L^2\left(\mathbb{R}^{(m,n)}, d\xi\right) \longrightarrow \mathcal{H}_{\mathcal{M},\alpha}$$

is an isometry.

Proof. We already showed that $\vartheta_{\mathcal{M},\alpha}$ preserves the norms. First we observe that if $(\lambda_0,\mu_0,\kappa_0) \in \Gamma_L$ and $g = [\lambda,\mu,\kappa] \in G$,

$$\begin{aligned}
(\lambda_0,\mu_0,\kappa_0)\circ g &= [\lambda_0,\mu_0,\kappa_0+\mu_0\,{}^t\lambda_0] \diamond [\lambda,\mu,\kappa] \\
&= [\lambda_0+\lambda,\mu_0+\mu,\kappa+\kappa_0+\mu_0\,{}^t\lambda_0+\lambda_0\,{}^t\mu+\mu\,{}^t\lambda_0].
\end{aligned}$$

Thus we get

$$
\begin{aligned}
&(\vartheta_{\mathcal{M},\alpha}f)((\lambda_0,\mu_0,\kappa_0)\circ g)\\
&= e^{2\pi i\,\sigma\{\mathcal{M}(\kappa+\kappa_0+\mu_0\,{}^t\lambda_0+\lambda_0\,{}^t\mu+\mu\,{}^t\lambda_0)\}}\\
&\quad\times\sum_{N\in\mathbb{Z}^{(m,n)}} f(\lambda+\lambda_0+N)\,e^{2\pi i\,\sigma\{(\alpha+2\mathcal{M}N)\,{}^t(\mu_0+\mu)\}}\\
&= e^{2\pi i\,\sigma(\mathcal{M}\kappa_0)}\cdot e^{2\pi i\,\sigma(\mathcal{M}\kappa)}\cdot e^{2\pi i\,\sigma(\alpha\,{}^t\mu_0)}\sum_{N\in\mathbb{Z}^{(m,n)}} f(\lambda+N)\,e^{2\pi i\,\sigma\{(\alpha+2\mathcal{M}N)\,{}^t\mu\}}\\
&= e^{2\pi i\,\sigma(\mathcal{M}\kappa_0)}\,(\vartheta_{\mathcal{M},\alpha}f)(g).
\end{aligned}
$$

Here in the above equalities we used the facts that $2\sigma(\mathcal{M}N\,{}^t\mu_0)\in\mathbb{Z}$ and $\alpha\,{}^t\mu_0\in\mathbb{Z}$. It is easy to show that

$$\int_{\Gamma_L\backslash G}|\vartheta_{\mathcal{M},\alpha}f(\bar g)|^2\,d\bar g=\int_{\mathbb{R}^{(m,n)}}|f(\lambda)|^2\,d\lambda=\|f\|_2^2<\infty.$$

This completes the proof of Lemma 18.

Finally it is easy to show that for each $\alpha\in\mathcal{J}$, the mapping $\vartheta_{\mathcal{M},\alpha}$ intertwines the Schrödinger representation $\left(U^{S,\mathcal{M}},L^2\left(\mathbb{R}^{(m,n)},d\xi\right)\right)$ and the representation $(\pi_{\mathcal{M},\alpha},\mathcal{H}_{\mathcal{M},\alpha})$. Therefore, by Lemma 18, for each $\alpha\in\mathcal{J}$, $\pi_{\mathcal{M},\alpha}$ is unitarily equivalent to $U(\sigma_{\mathcal{M}})$ and so $\pi_{\mathcal{M},\alpha}$ is an irreducible unitary representation of G. According to (1.6.20), the induced representation $\pi_{\mathcal{M}}$ is unitarily equivalent to

$$\bigoplus U_{\mathcal{M}}\quad\left(\,(\det 2\mathcal{M})^n\text{-copies}\,\right).$$

This completes the proof of Theorem 8. □

Now we state the connection between the lattice representation and theta functions. As before, we write $V=\mathbb{R}^{(m,n)}\times\mathbb{R}^{(m,n)}\cong\mathbb{C}^{(m,n)}$, $L=\mathbb{Z}^{(m,n)}\times\mathbb{Z}^{(m,n)}$ and $\mathcal{M}$ is a positive symmetric half-integral matrix of degree m. The function $q_{\mathcal{M}}:L\longrightarrow\mathbb{R}/2\mathbb{Z}=[0,2)$ defined by

$$q_{\mathcal{M}}((\xi,\eta))=2\,\sigma(\mathcal{M}\xi\,{}^t\eta),\quad(\xi,\eta)\in L\tag{1.6.24}$$

satisfies the condition (1.6.8). We let $\varphi_{\mathcal{M},q_{\mathcal{M}}}:\Gamma_L\longrightarrow\mathbb{C}_1^*$ be the character of Γ_L defined by

$$\varphi_{\mathcal{M},q_{\mathcal{M}}}((l,\kappa))=e^{2\pi i\,\sigma(\mathcal{M}\kappa)}\,e^{\pi i\,q_{\mathcal{M}}(l)},\quad(l,\kappa)\in\Gamma_L.$$

We denote by $\mathcal{H}_{\mathcal{M},q_{\mathcal{M}}}$ the Hilbert space consisting of measurable functions $\phi:G\longrightarrow\mathbb{C}$ which satisfy the conditions (1.6.24) and (1.6.25):

$$\phi((l,\kappa)\circ g)=\varphi_{\mathcal{M},q_{\mathcal{M}}}((l,\kappa))\,\phi(g)\ \text{ for all }(l,\kappa)\in\Gamma_L\text{and }g\in G.\tag{1.6.25}$$

$$\int_{\Gamma_L \backslash G} \|\phi(\dot{g})\|^2 \, d\dot{g} \, l \, \infty, \quad \dot{g} = \Gamma_L \circ g. \tag{1.6.26}$$

Then the representation

$$\pi_{\mathcal{M},q_{\mathcal{M}}} = \mathrm{Ind}_{\Gamma_L}^G \varphi_{\mathcal{M},q_{\mathcal{M}}}$$

of G induced from the character $\varphi_{\mathcal{M},q_{\mathcal{M}}}$ is realized in $\mathcal{H}_{\mathcal{M},q_{\mathcal{M}}}$ as

$$(\pi_{\mathcal{M},q_{\mathcal{M}}}(g_0)\,\phi)(g) = \phi(gg_0), \quad g_0, g \in G, \ \phi \in \mathcal{H}_{\mathcal{M},q_{\mathcal{M}}}.$$

Let $\mathbf{H}_{\mathcal{M},q_{\mathcal{M}}}$ be the vector space consisting of measurable functions $F : V \longrightarrow \mathbb{C}$ satisfying the conditions (1.6.26) and (1.6.27).

$$F(\lambda+\xi, \mu+\eta) = e^{2\pi i\,\sigma\{\mathcal{M}(\xi\,{}^t\eta + \lambda\,{}^t\eta - \mu\,{}^t\xi)\}}\, F(\lambda,\mu) \tag{1.6.27}$$

for all $(\lambda,\mu) \in V$ and $(\xi,\eta) \in L$.

$$\int_{L\backslash V} \|F(\dot{v})\|^2 \, d\dot{g} = \int_{I_\lambda \times I_\mu} \|F(\lambda,\mu)\|^2 \, d\lambda d\mu \, l \, \infty. \tag{1.6.28}$$

Given $\phi \in \mathcal{H}_{\mathcal{M},q_{\mathcal{M}}}$ and a fixed element $\Omega \in \mathbb{H}_n$, we put

$$E_\phi(\lambda,\mu) = \phi((\lambda,\mu,0)), \quad \lambda,\mu \in \mathbb{R}^{(m,n)}, \tag{1.6.29}$$

$$F_\phi(\lambda,\mu) = \phi([\lambda,\mu,0]), \quad \lambda,\mu \in \mathbb{R}^{(m,n)}, \tag{1.6.30}$$

$$F_{\Omega,\phi}(\lambda,\mu) = e^{-2\pi i\,\sigma(\mathcal{M}\lambda\Omega\,{}^t\lambda)}\, F_\phi(\lambda,\mu), \quad \lambda,\mu \in \mathbb{R}^{(m,n)}. \tag{1.6.31}$$

In addition, we put for $W = \lambda\Omega + \mu \in \mathbb{C}^{(m,n)}$,

$$\vartheta_{\Omega,\phi}(W) = \vartheta_{\Omega,\phi}(\lambda\Omega+\mu) := F_{\Omega,\phi}(\lambda,\mu). \tag{1.6.32}$$

We observe that E_ϕ, F_ϕ, $F_{\Omega,\phi}$ are functions defined on V and $\vartheta_{\Omega,\phi}$ is a function defined on $\mathbb{C}^{(m,n)}$.

Proposition 6. *If $\phi \in \mathcal{H}_{\mathcal{M},q_{\mathcal{M}}}$, $(\xi,\eta) \in L$ and $(\lambda,\mu) \in V$, then we have the formulas*

$$E_\phi(\lambda+\xi, \mu+\eta) = e^{2\pi i\,\sigma\{\mathcal{M}(\xi\,{}^t\eta + \lambda\,{}^t\eta - \mu\,{}^t\xi)\}}\, E_\phi(\lambda,\mu). \tag{1.6.33}$$

$$F_\phi(\lambda+\xi, \mu+\eta) = e^{-4\pi i\,\sigma(\mathcal{M}\xi\,{}^t\mu)}\, F_\phi(\lambda,\mu). \tag{1.6.34}$$

$$F_{\Omega,\phi}(\lambda+\xi, \mu+\eta) = e^{-2\pi i\,\sigma\{\mathcal{M}(\xi\Omega\,{}^t\xi + 2\,\lambda\Omega\,{}^t\xi + 2\,\mu\,{}^t\xi)\}}\, F_{\Omega,\phi}(\lambda,\mu). \tag{1.6.35}$$

If $W = \lambda\Omega + \eta \in \mathbb{C}^{(m,n)}$, then we have

$$\vartheta_{\Omega,\phi}(W + \xi\Omega + \eta) = e^{-2\pi i\,\sigma\{\mathcal{M}(\xi\Omega\,{}^t\xi + 2\,W\,{}^t\xi)\}}\, \vartheta_{\Omega,\phi}(W). \tag{1.6.36}$$

Moreover, F_ϕ is an element of $\mathbf{H}_{\mathcal{M},q_{\mathcal{M}}}$.

Proof. We note that

$$(\lambda+\xi,\mu+\eta,0)=(\xi,\eta,-\xi\,{}^t\mu+\eta\,{}^t\lambda)\circ(\lambda,\mu,0).$$

Thus we have

$$\begin{aligned}E_\phi(\lambda+\xi,\mu+\eta)&=\phi((\lambda+\xi,\mu+\eta,0))\\&=\phi((\xi,\eta,\,-\xi\,{}^t\mu+\eta\,{}^t\lambda)\circ(\lambda,\mu,0))\\&=e^{2\pi i\,\sigma\{\mathcal{M}(\xi\,{}^t\eta+\lambda\,{}^t\eta-\mu\,{}^t\xi)\}}\,\phi((\lambda,\mu,0))\\&=e^{2\pi i\,\sigma\{\mathcal{M}(\xi\,{}^t\eta+\lambda\,{}^t\eta-\mu\,{}^t\xi)\}}\,E_\phi(\lambda,\mu).\end{aligned}$$

This proves the formula (1.6.33).

We observe that

$$[\lambda+\xi,\mu+\eta,0]=(\xi,\eta,\,-\xi\,{}^t\mu-\mu\,{}^t\xi-\eta\,{}^t\xi)\circ[\lambda,\mu,0].$$

Thus we have

$$\begin{aligned}F_\phi(\lambda+\xi,\mu+\eta)&=\phi([\lambda+\xi,\mu+\eta,0])\\&=e^{-2\pi i\,\sigma\{\mathcal{M}(\xi\,{}^t\mu+\mu\,{}^t\xi+\eta\,{}^t\xi)\}}\\&\quad\times e^{2\pi i\,\sigma(\mathcal{M}\xi\,{}^t\eta)}\,\phi([\lambda,\mu,0])\\&=e^{-4\pi i\,\sigma(\mathcal{M}\xi\,{}^t\mu)}\,\phi([\lambda,\mu,0])\\&=e^{-4\pi i\,\sigma(\mathcal{M}\xi\,{}^t\mu)}\,F_\phi(\lambda,\mu).\end{aligned}$$

This proves the formula (1.6.34).

According to (1.6.34), we have

$$\begin{aligned}F_{\Omega,\phi}(\lambda+\xi,\mu+\eta)&=e^{-2\pi i\,\sigma\{\mathcal{M}(\lambda+\xi)\Omega\,{}^t(\lambda+\xi)\}}\,F_\phi(\lambda+\xi,\mu+\eta)\\&=e^{-2\pi i\,\sigma\{\mathcal{M}(\lambda+\xi)\Omega\,{}^t(\lambda+\xi)\}}\\&\quad\times e^{-4\pi i\,\sigma(\mathcal{M}\xi\,{}^t\mu)}\,F_\phi(\lambda,\mu)\\&=e^{-2\pi i\,\sigma\{\mathcal{M}(\xi\Omega\,{}^t\xi+2\,\lambda\Omega\,{}^t\xi+2\,\mu\,{}^t\xi)\}}\\&\quad\times e^{-2\pi i\,\sigma(\mathcal{M}\lambda\Omega\,{}^t\lambda)}\,F_\phi(\lambda,\mu)\\&=e^{-2\pi i\,\sigma\{\mathcal{M}(\xi\Omega\,{}^t\xi+2\,\lambda\Omega\,{}^t\xi+2\,\mu\,{}^t\xi)\}}\,F_{\Omega,\phi}(\lambda,\mu).\end{aligned}$$

This proves the formula (1.6.34). The formula (1.6.35) follows immediately from the formula (1.6.34).

Indeed, if $W=\lambda\Omega+\mu$ with $\lambda,\mu\in\mathbb{R}^{(m,n)}$, we have

$$\begin{aligned}\vartheta_{\Omega,\phi}(W+\xi\Omega+\eta) &= F_{\Omega,\phi}(\lambda+\xi,\mu+\eta)\\ &= e^{-2\pi i\,\sigma\{\mathcal{M}(\xi\Omega\,{}^t\xi\,+2\,(\lambda\Omega+\mu)\,{}^t\xi)\}}\,F_{\Omega,\phi}(\lambda,\mu)\\ &= e^{-2\pi i\,\sigma\{\mathcal{M}(\xi\Omega\,{}^t\xi\,+2\,W\,{}^t\xi)\}}\,\vartheta_{\Omega,\phi}(W).\end{aligned}$$

□

Remark 4. The function $\vartheta_{\Omega,\phi}(W)$ is a theta function of level $2\mathcal{M}$ with respect to Ω if $\vartheta_{\Omega,\phi}$ is holomorphic. For any $\phi\in\mathcal{H}_{\mathcal{M},q_{\mathcal{M}}}$, the function $\vartheta_{\Omega,\phi}$ satisfies the transformation law (1.2.1) of a theta function. In this sense, the lattice representation $(\pi_{\mathcal{M},q_{\mathcal{M}}},\mathcal{H}_{\mathcal{M},q_{\mathcal{M}}})$ is closely related to

1.7 Coadjoint Orbits

In this section, we find the coadjoint orbits of the Heisenberg group $H_{\mathbb{R}}^{(n,m)}$ and describe the connection between the coadjoint orbits and the unitary dual of $H_{\mathbb{R}}^{(n,m)}$ explicitly.

For brevity, we let $G := H_{\mathbb{R}}^{(n,m)}$ as before. Let $\mathfrak{g}$ be the Lie algebra of G and let $\mathfrak{g}^*$ be the dual space of $\mathfrak{g}$. We observe that $\mathfrak{g}$ can be regarded as the subalgebra consisting of all $(m+n) \times (m+n)$ real matrices of the form

$$X(\alpha,\beta,\gamma) := \begin{pmatrix} 0 & 0 & 0 & {}^t\beta \\ \alpha & 0 & \beta & \gamma \\ 0 & 0 & 0 & -{}^t\alpha \\ 0 & 0 & 0 & 0 \end{pmatrix}, \ \alpha,\beta \in \mathbb{R}^{(m,n)}, \ \gamma = {}^t\gamma \in \mathbb{R}^{(m,m)}$$

of the Lie algebra $\mathfrak{sp}(m+n,\mathbb{R})$ of the symplectic group $Sp(m+n,\mathbb{R})$. An easy computation yields

$$[X(\alpha,\beta,\gamma),\, X(\delta,\epsilon,\xi)] = X(0,0,\alpha\,{}^t\epsilon + \epsilon\,{}^t\alpha - \beta\,{}^t\delta - \delta\,{}^t\beta).$$

The dual space $\mathfrak{g}^*$ of $\mathfrak{g}$ can be identified with the vector space consisting of all $(m+n) \times (m+n)$ real matrices of the form

$$F(a,b,c) := \begin{pmatrix} 0 & {}^ta & 0 & 0 \\ 0 & 0 & 0 & 0 \\ 0 & {}^tb & 0 & 0 \\ b & c & -a & 0 \end{pmatrix}, \ a,b \in \mathbb{R}^{(m,n)}, \ c = {}^tc \in \mathbb{R}^{(m,m)}$$

so that

$$\begin{aligned} \langle F(a,b,c), X(\alpha,\beta,\gamma)\rangle :&= \sigma(F(a,b,c)\,X(\alpha,\beta,\gamma)) \\ &= 2\,\sigma({}^t\alpha\, a + {}^tb\,\beta) + \sigma(c\,\gamma). \end{aligned} \tag{1.7.1}$$

The adjoint representation Ad of G is given by $Ad_G(g)X = gXg^{-1}$ for $g \in G$ and $X \in \mathfrak{g}$. For $g \in G$ and $F \in \mathfrak{g}^*$, gFg^{-1} is not of the form $F(a,b,c)$. We denote by $(gFg^{-1})_*$ the

$$\begin{pmatrix} 0 & * & 0 & 0 \\ 0 & 0 & 0 & 0 \\ 0 & * & 0 & 0 \\ * & * & * & 0 \end{pmatrix} - \text{part}$$

of the matrix gFg^{-1}. Then it is easy to see that the coadjoint representation $Ad_G^* : G \longrightarrow GL(\mathfrak{g}^*)$ is given by $Ad_G^*(g)F = (gFg^{-1})_*$, where $g \in G$ and $F \in \mathfrak{g}^*$. More precisely,

$$Ad_G^*(g)F(a,b,c) = F(a + c\mu, b - c\lambda, c), \tag{1.7.2}$$

where $g = (\lambda, \mu, \kappa) \in G$. Thus the coadjoint orbit $\Omega_{a,b}$ of G at $F(a,b,0) \in \mathfrak{g}^*$ is given by

$$\Omega_{a,b} = Ad_G^*(G)\, F(a,b,0) = \{F(a,b,0)\}, \text{ a single point} \tag{1.7.3}$$

and the coadjoint orbit Ω_c of G at $F(0,0,c) \in \mathfrak{g}^*$ with $c \neq 0$ is given by

$$\Omega_c = Ad_G^*(G)\, F(0,0,c) = \left\{ F(a,b,c) \,\middle|\, a,b \in \mathbb{R}^{(m,n)} \right\}. \tag{1.7.4}$$

Therefore the coadjoint orbits of G in $\mathfrak{g}^*$ fall into two classes:

(I) The single point $\left\{ \Omega_{a,b} \,\middle|\, a,b \in \mathbb{R}^{(m,n)} \right\}$ located in the plane $c = 0$.
(II) The affine planes $\left\{ \Omega_c \,\middle|\, c = {}^t c \in \mathbb{R}^{(m,m)},\ c \neq 0 \right\}$ parallel to the homogeneous plane $c = 0$.

In other words, the orbit space $\mathcal{O}(G)$ of coadjoint orbits is parametrized by

$$\begin{cases} c-\text{axis},\ c \neq 0,\ c = {}^t c \in \mathbb{R}^{(m,m)}; \\ (a,b)-\text{plane} \cong \mathbb{R}^{(m,n)} \times \mathbb{R}^{(m,n)}. \end{cases}$$

The single point coadjoint orbits of the type $\Omega_{a,b}$ are said to be the **degenerate** orbits of G in $\mathfrak{g}^*$. On the other hand, the flat coadjoint orbits of the type Ω_c are said to be the **non-degenerate** orbits of G in $\mathfrak{g}^*$.

Since G is connected and simply connected 2-step nilpotent Lie group, according to A. Kirillov (cf. [16] or [17] p.249, Theorem 1), the unitary dual $\widehat{G}$ of G is given by

$$\widehat{G} = \left(\mathbb{R}^{(m,n)} \times \mathbb{R}^{(m,n)} \right) \coprod \left\{ z \in \mathbb{R}^{(m,m)} \mid z = {}^t z,\ z \neq 0 \right\}, \tag{1.7.5}$$

where $\coprod$ denotes the disjoint union. The topology of $\widehat{G}$ may be described as follows. The topology on $\{c\text{-axis} - (0)\}$ is the usual topology of the Euclidean space and the topology on $\{F(a,b,0) | a,b \in \mathbb{R}^{(m,n)}\}$ is the usual Euclidean topology. But a sequence on the c-axis which converges to 0 in the usual topology converges to the whole Euclidean space $\mathbb{R}^{(m,n)} \times \mathbb{R}^{(m,n)}$ in the topology of $\widehat{G}$. This is just the quotient topology on $\mathfrak{g}^*/G$ so that algebraically and topologically $\widehat{G} = \mathfrak{g}^*/G$.

It is well known that each coadjoint orbit is a symplectic manifold. We will state this fact in detail. For the present time being, we fix an element F of $\mathfrak{g}^*$ once and for all. We consider the alternating $\mathbb{R}$-bilinear form $\mathbf{B}_F$ on $\mathfrak{g}$ defined by

$$\mathbf{B}_F(X,Y) := \langle F, [X,Y] \rangle = \langle ad_{\mathfrak{g}}^*(Y)F, X \rangle, \quad X, Y \in \mathfrak{g}, \tag{1.7.6}$$

where $ad_{\mathfrak{g}}^* : \mathfrak{g} \longrightarrow \mathrm{End}(\mathfrak{g}^*)$ denotes the differential of the coadjoint representation $Ad_G^* : G \longrightarrow GL(\mathfrak{g}^*)$. More precisely, if $F = F(a,b,c),\ X =$

$X(\alpha,\beta,\gamma)$, and $Y = X(\delta,\epsilon,\xi)$, then

$$\mathbf{B}_F(X,Y) = \sigma\{c\,(\alpha\,{}^t\epsilon + \epsilon\,{}^t\alpha - \beta\,{}^t\delta - \delta\,{}^t\beta)\}. \tag{1.7.7}$$

For $F \in \mathfrak{g}^*$, we let

$$G_F = \{\, g \in G \mid Ad_G^*(g)F = F\,\}$$

be the stabilizer of the coadjoint action Ad^* of G on $\mathfrak{g}^*$ at F. Since G_F is a closed subgroup of G, G_F is a Lie subgroup of G. We denote by $\mathfrak{g}_F$ the Lie subalgebra of $\mathfrak{g}$ corresponding to G_F. Then it is easy to show that

$$\mathfrak{g}_F = \operatorname{rad}\mathbf{B}_F = \left\{\, X \in \mathfrak{g} \mid ad_{\mathfrak{g}}^*(X)F = 0\,\right\}. \tag{1.7.8}$$

Here $\operatorname{rad}\mathbf{B}_F$ denotes the radical of $\mathbf{B}_F$ in $\mathfrak{g}$. We let $\dot{\mathbf{B}}_F$ be the non-degenerate alternating $\mathbb{R}$-bilinear form on the quotient vector space $\mathfrak{g}/\mathrm{rad}\ \mathbf{B}_F$ induced from $\mathbf{B}_F$. Since we may identify the tangent space of the coadjoint orbit $\Omega_F \cong G/G_F$ with $\mathfrak{g}/\mathfrak{g}_F = \mathfrak{g}/\mathrm{rad}\,\mathbf{B}_F$, we see that the tangent space of Ω_F at F is a symplectic vector space with respect to the symplectic form $\dot{\mathbf{B}}_F$.

Now we are ready to prove that the coadjoint orbit $\Omega_F = Ad_G^*(G)F$ is a symplectic manifold. We denote by $\widetilde{X}$ the smooth vector field on $\mathfrak{g}^*$ associated to $X \in \mathfrak{g}$. That means that for each $\ell \in \mathfrak{g}^*$, we have

$$\widetilde{X}(\ell) = ad_{\mathfrak{g}}^*(X)\,\ell. \tag{1.7.9}$$

We define the differential 2-form B_{Ω_F} on Ω_F by

$$B_{\Omega_F}(\widetilde{X},\widetilde{Y}) = B_{\Omega_F}(ad_{\mathfrak{g}}^*(X)F, ad_{\mathfrak{g}}^*(Y)F) := \mathbf{B}_F(X,Y), \tag{1.7.10}$$

where $X, Y \in \mathfrak{g}$.

Lemma 19. *B_{Ω_F} is non-degenerate.*

Proof. Let $\widetilde{X}$ be the smooth vector field on $\mathfrak{g}^*$ associated to $X \in \mathfrak{g}$ such that $B_{\Omega_F}(\widetilde{X},\widetilde{Y}) = 0$ for all $\widetilde{Y}$ with $Y \in \mathfrak{g}$. Since $B_{\Omega_F}(\widetilde{X},\widetilde{Y}) = \mathbf{B}_F(X,Y) = 0$ for all $Y \in \mathfrak{g}$, $X \in \mathfrak{g}_F$. Thus $\widetilde{X} = 0$. Hence B_{Ω_F} is non-degenerate. □

Lemma 20. *B_{Ω_F} is closed.*

Proof. If $\widetilde{X_1}$, $\widetilde{X_2}$, and $\widetilde{X_3}$ are three smooth vector fields on $\mathfrak{g}^*$ associated to $X_1, X_2, X_3 \in \mathfrak{g}$, then

$$\begin{aligned}
&dB_{\Omega_F}(\widetilde{X_1},\widetilde{X_2},\widetilde{X_3})\\
&= \widetilde{X_1}(B_{\Omega_F}(\widetilde{X_2},\widetilde{X_3})) - \widetilde{X_2}(B_{\Omega_F}(\widetilde{X_1},\widetilde{X_3})) + \widetilde{X_3}(B_{\Omega_F}(\widetilde{X_1},\widetilde{X_2}))\\
&\quad - B_{\Omega_F}([\widetilde{X_1},\widetilde{X_2}],\widetilde{X_3}) + B_{\Omega_F}([\widetilde{X_1},\widetilde{X_3}],\widetilde{X_2}) - B_{\Omega_F}([\widetilde{X_2},\widetilde{X_3}],\widetilde{X_1})\\
&= -\langle F, [[X_1,X_2],X_3] + [[X_2,X_3],X_1] + [[X_3,X_1],X_2]\rangle\\
&= 0 \qquad \text{(by the Jacobi identity).}
\end{aligned}$$

Therefore B_{Ω_F} is closed. □

In summary, (Ω_F, B_{Ω_F}) is a symplectic manifold of dimension $2mn$ or 0.

In order to describe the irreducible unitary representations of G corresponding to the coadjoint orbits under the Kirillov correspondence, we have to determine the polarizations of $\mathfrak{g}$ for the linear forms $F \in \mathfrak{g}^*$.

Case I. $F = F(a, b, 0)$; the degenerate case.

According to (1.7.3), $\Omega_F = \Omega_{a,b} = \{F(a, b, 0)\}$ is a single point. It follows from (1.7.7) that $\mathbf{B}_F(X, Y) = 0$ for all $X, Y \in \mathfrak{g}$. Thus $\mathfrak{g}$ is the unique polarization of $\mathfrak{g}$ for F. The Kirillov correspondence says that the irreducible unitary representation $\pi_{a,b}$ of G corresponding to the coadjoint orbit $\Omega_{a,b}$ is given by

$$\pi_{a,b}(\exp X(\alpha, \beta, \gamma)) = e^{2\pi i \langle F, X(\alpha,\beta,\gamma)\rangle} = e^{4\pi i\, \sigma({}^t a\alpha + {}^t b\beta)}. \tag{1.7.11}$$

That is, $\pi_{a,b}$ is a one-dimensional degenerate representation of G.

Case II. $F = F(0, 0, c)$, $0 \neq c = {}^t c \in \mathbb{R}^{(m,m)}$: the non-degenerate case.

According to (1.7.4), $\Omega_F = \Omega_c = \left\{ F(a, b, c) \mid a, b \in \mathbb{R}^{(m,n)} \right\}$. By (1.7.7), we see that

$$\mathfrak{k} = \left\{ X(0, \beta, \gamma) \mid \beta \in \mathbb{R}^{(m,n)},\ \gamma = {}^t\gamma \in \mathbb{R}^{(m,m)} \right\} \tag{1.7.12}$$

is a polarization of $\mathfrak{g}$ for F, i.e.,$\mathfrak{k}$ is a Lie subalgebra of $\mathfrak{g}$ subordinate to $F \in \mathfrak{g}^*$ which is maximal among the totally isotropic vector subspaces of $\mathfrak{g}$ relative to the alternating $\mathbb{R}$-bilinear form $\mathbf{B}_F$. Let K be the simply connected Lie subgroup of G corresponding to the Lie subalgebra $\mathfrak{k}$ of $\mathfrak{g}$. We let

$$\chi_{c,\mathfrak{k}} : K \longrightarrow \mathbb{C}_1^\times$$

be the unitary character of K defined by

$$\chi_{c,\mathfrak{k}}(\exp X(0, \beta, \gamma)) = e^{2\pi i \langle F, X(0,\beta,\gamma)\rangle} = e^{2\pi i\, \sigma(c\gamma)}. \tag{1.7.13}$$

The Kirillov correspondence says that the irreducible unitary representation $\pi_{c,\mathfrak{k}}$ of G corresponding to the coadjoint orbit $\Omega_F = \Omega_c$ is given by

$$\pi_{c,\mathfrak{k}} = \mathrm{Ind}_K^G\, \chi_{c,\mathfrak{k}}. \tag{1.7.14}$$

According to Kirillov's Theorem (cf. [16]), we know that the induced representation $\pi_{c,\mathfrak{k}}$ is, up to equivalence, independent of the choice of a polarization of $\mathfrak{g}$ for F. Thus we denote the equivalence class of $\pi_{c,\mathfrak{k}}$ by π_c. π_c is realized on the representation space $L^2(\mathbb{R}^{(m,n)}, d\xi)$ as follows:

$$(\pi_c(g)f)(\xi) = e^{2\pi i\, \sigma\{c(\kappa + \mu\, {}^t\lambda + 2\xi\, {}^t\mu)\}} f(\xi + \lambda), \tag{1.7.15}$$

where $g = (\lambda, \mu, \kappa) \in G$ and $\xi \in \mathbb{R}^{(m,n)}$. Using the fact that

$$\exp X(\alpha,\beta,\gamma) = \left(\alpha,\beta,\gamma + \frac{1}{2}\left(\alpha\,{}^t\beta - \beta\,{}^t\alpha\right)\right),$$

we see that π_c is nothing but the Schrödinger representation $U_c = U(\sigma_c)$ of G induced from the one-dimensional unitary representation σ_c of K given by $\sigma_c((0,\mu,\kappa)) = e^{2\pi i\,\sigma(c\kappa)}I$ (cf. (1.4.6) and (1.4.8)). We note that π_c is the non-degenerate representation of G with central character $\chi_c : Z \longrightarrow \mathbb{C}_1^\times$ given by $\chi_c((0,0,\kappa)) = e^{2\pi i\,\sigma(c\kappa)}$. Here $Z = \left\{(0,0,\kappa) \mid \kappa = {}^t\kappa \in \mathbb{R}^{(m,m)}\right\}$ denotes the center of G.

It is well known that the monomial representation $\left(\pi_c, L^2\left(\mathbb{R}^{(m,n)}, d\xi\right)\right)$ of G extends to an operator of trace class

$$\pi_c(\phi) : L^2\left(\mathbb{R}^{(m,n)}, d\xi\right) \longrightarrow L^2\left(\mathbb{R}^{(m,n)}, d\xi\right) \tag{1.7.16}$$

for all $\phi \in C_c^\infty(G)$. Here $C_c^\infty(G)$ is the vector space of all smooth functions on G with compact support. We let $C_c^\infty(\mathfrak{g})$ and $C(\mathfrak{g}^*)$ the vector space of all smooth functions on $\mathfrak{g}$ with compact support and the vector space of all continuous functions on $\mathfrak{g}^*$ respectively. If $f \in C_c^\infty(\mathfrak{g})$, we define the Fourier cotransform

$$\mathcal{C}F_{\mathfrak{g}} : C_c^\infty(\mathfrak{g}) \longrightarrow C(\mathfrak{g}^*)$$

by

$$\left(\mathcal{C}F_{\mathfrak{g}}(f)\right)(F') := \int_{\mathfrak{g}} f(X)\, e^{2\pi i\,\langle F', X\rangle} dX, \tag{1.7.17}$$

where $F' \in \mathfrak{g}^*$ and dX denotes the usual Lebesgue measure on $\mathfrak{g}$. According to A. Kirillov (cf. [16]), there exists a measure β on the coadjoint orbit $\Omega_c \approx \mathbb{R}^{(m,n)} \times \mathbb{R}^{(m,n)}$ which is invariant under the coadjoint action of G such that

$$\operatorname{tr}\pi_c^1(\phi) = \int_{\Omega_c} \mathcal{C}F_{\mathfrak{g}}(\phi\circ\exp)(F')d\beta(F') \tag{1.7.18}$$

holds for all test functions $\phi \in C_c^\infty(G)$, where exp denotes the exponentional mapping of $\mathfrak{g}$ onto G. We recall that

$$\pi_c^1(\phi)(f) := \int_G \phi(x)\left(\pi_c(x)f\right)dx,$$

where $\phi \in C_c^\infty(G)$ and $f \in L^2(\mathbb{R}^{(m,n)}, d\xi)$. By the Plancherel theorem, the mapping

$$S(G/Z) \ni \varphi \longmapsto \pi_c^1(\varphi) \in TC(L^2(\mathbb{R}^{(m,n)}, d\xi))$$

extends to a unitary isometry

$$\pi_c^2 : L^2(G/Z, \chi_c) \longrightarrow HS\left(L^2\left(\mathbb{R}^{(m,n)}, d\xi\right)\right) \tag{1.7.19}$$

of the representation space $L^2(G/Z, \chi_c)$ of $\operatorname{Ind}_Z^G \chi_c$ onto the complex Hilbert space $HS\left(L^2\left(\mathbb{R}^{(m,n)}, d\xi\right)\right)$ consisting of all Hilbert-Schmidt operators on

$L^2\big(\mathbb{R}^{(m,n)}, d\xi\big)$, where $S(G/Z)$ is the Schwartz space of all infinitely differentiable complex-valued functions on $G/Z \cong \mathbb{R}^{(m,n)} \times \mathbb{R}^{(m,n)}$ that are rapidly decreasing at infinity and $TC\big(L^2\big(\mathbb{R}^{(m,n)}, d\xi\big)\big)$ denotes the complex vector space of all continuous $\mathbb{C}$-linear mappings of $L^2\big(\mathbb{R}^{(m,n)}, d\xi\big)$ into itself which are of trace class.

In summary, we have the following result.

Theorem 9. *For $F = F(a,b,0) \in \mathfrak{g}^*$, the irreducible unitary representation $\pi_{a,b}$ of G corresponding to the coadjoint orbit $\Omega_F = \Omega_c$ under the Kirillov correspondence is degenerate representation of G given by*

$$\pi_{a,b}\big(\exp X(\alpha,\beta,\gamma)\big) = e^{4\pi i\,\sigma({}^t a\alpha - {}^t b\beta)}.$$

On the other hand, for $F = F(0,0,c) \in \mathfrak{g}^*$ with $0 \neq c = {}^t c \in \mathbb{R}^{(m,m)}$, the irreducible unitary representation $\big(\pi_c, L^2\big(\mathbb{R}^{(m,n)}, d\xi\big)\big)$ of G corresponding to the coadjoint orbit Ω_c under the Kirillov correspondence is unitary equivalent to the Schrödinger representation $\big(U_c, L^2\big(\mathbb{R}^{(m,n)}, d\xi\big)\big)$ and this non-degenerate representation π_c is square integrable modulo its center Z. For all test functions $\phi \in C_c^\infty(G)$, the character formula

$$tr\,\pi_c^2(\phi) = \mathcal{C}(\phi, c) \int_{\mathbb{R}^{(m,n)}} \phi(0,0,\kappa)\, e^{2\pi i\,\sigma(c\kappa)} d\kappa$$

holds for some constant $\mathcal{C}(\phi,c)$ depending on ϕ and c, where $d\kappa$ is the Lebesgue measure on the Euclidean space $\mathbb{R}^{(m,m)}$.

Now we consider the subgroup K of G (cf. (1.4.1)) given by

$$K := \big\{\,(0,0,\kappa) \in G \,\big|\, \mu \in \mathbb{R}^{(m,n)},\ \kappa = {}^t\kappa \in \mathbb{R}^{(m,m)}\,\big\}.$$

The Lie algebra $\mathfrak{k}$ of K is given by (1.7.12). The dual space $\mathfrak{k}^*$ of $\mathfrak{k}$ may be identified with the space

$$\big\{F(0,b,c) \,\big|\, b \in \mathbb{R}^{(m,n)},\ c = {}^t c \in \mathbb{R}^{(m,m)}\big\}.$$

We let $Ad_K^* : K \longrightarrow GL(\mathfrak{k}^*)$ be the coadjoint representation of K on $\mathfrak{k}^*$. The coadjoint orbit $\omega_{b,c}$ of K at $F(0,b,c) \in \mathfrak{k}^*$ is given by

$$\omega_{b,c} = Ad_K^*(K)\, F(0,b,c) = \{F(0,b,c)\},\ \text{ a single point.} \tag{1.7.20}$$

Since K is a commutative group, $[\mathfrak{k},\mathfrak{k}] = 0$ and so the alternating $\mathbb{R}$-bilinear form $\mathbf{B}_f$ on $\mathfrak{k}$ associated to $F := F(0,b,c)$ identically vanishes on $\mathfrak{k} \times \mathfrak{k}$(cf. (1.7.6)). $\mathfrak{k}$ is the unique polarization of $\mathfrak{k}$ for $F = F(0,b,c)$. The Kirillov correspondence says that the irreducible unitary representation $\chi_{b,c}$ of K corresponding to the coadjoint orbit $\omega_{b,c}$ is given by

$$\chi_{b,c}\big(\exp X(0,\beta,\gamma)\big) = e^{2\pi i\,\langle F(0,b,c), X(0,\beta,\gamma)\rangle} = e^{2\pi i\,\sigma(2\,{}^t b\,\beta + c\,\gamma)} \tag{1.7.21}$$

or

$$\chi_{b,c}((0,\mu,\kappa)) = e^{2\pi i\,\sigma(2\,{}^t b\,\mu + c\,\kappa)}, \quad (0,\mu,\kappa) \in K. \tag{1.7.22}$$

For $0 \neq c = {}^t c \in \mathbb{R}^{(m,m)}$, we let π_c be the Schrödinger representation of G given by (1.7.15). We know that the irreducible unitary representation of G corresponding to the coadjoint orbit

$$\Omega_c = \mathrm{Ad}_G^*(G)\, F(0,0,c) = \left\{ F(a,b,c) \,|\, a,b \in \mathbb{R}^{(m,n)} \right\}.$$

Let $p : \mathfrak{g}^* \longrightarrow \mathfrak{k}^*$ be the natural projection defined by $p(F(a,b,c)) = F(0,b,c)$. Obviously we have

$$p(\Omega_c) = \left\{ F(0,b,c) \,\middle|\, b \in \mathbb{R}^{(m,n)} \right\} = \bigcup_{b\in\mathbb{R}^{(m,n)}} \omega_{b,c}.$$

According to Kirillov Theorem (cf. [17] p.249, Theorem 1), the restriction $\pi_c|_K$ of π_c to K is the direct integral of all one-dimensional representations $\chi_{b,c}$ of K ($b \in \mathbb{R}^{(m,n)}$). Conversely, we let $\chi_{b,c}$ be the element of $\widehat{K}$ corresponding to the coadjoint orbit $\omega_{b,c}$ of K. The induced representation $\mathrm{Ind}_K^G\, \chi_{b,c}$ is nothing but the Schrödinger representation π_c. The coadjoint orbit Ω_c of G is the only coadjoint orbit such that $\Omega_c \cap p^{-1}(\omega_{b,c})$ is nonempty.

1.8 Hermite Operators

We recall the Schrödinger representation U_c of G induced from σ_c (cf. (5.8)). We consider the special case when $c = I_m$ is the identity matrix of degree m. Then it is easy to see that

$$\begin{aligned} dU_{I_m}(D^0_{kl})\, f(\xi) &= 2\,\pi\, i\, \delta_{kl}\, f(\xi), \\ dU_{I_m}(D_{ka})\, f(\xi) &= \frac{\partial f(\xi)}{\partial \xi_{ka}}, \\ dU_{I_m}(\widehat{D}_{lb})\, f(\xi) &= 4\,\pi\, i\, \xi_{lb}\, f(\xi), \end{aligned}$$

where $f \in \mathcal{S}(\mathbb{R}^{(m,n)})$ or $C^\infty(\mathbb{R})$, the Schwartz space and $\xi_{11}, \cdots, \xi_{mn}$ are the coordinates of ξ. In section two, we put

$$\begin{aligned} Z^0_{kl} &:= -i\, D^0_{kl}, \qquad 1 \le k \le l \le m, \\ Y^+_{ka} &:= \frac{1}{2}\,(D_{ka} + i\, \widehat{D}_{ka}), \quad 1 \le k \le m,\ 1 \le a \le n, \\ Y^-_{lb} &:= \frac{1}{2}\,(D_{lb} - i\, \widehat{D}_{lb}), \quad 1 \le l \le m,\ 1 \le b \le n. \end{aligned}$$

We set

$$A^+_{ka} := dU_{I_m}(Y^+_{ka}) = \frac{1}{2}\, dU_{I_m}(D_{ka}) + \frac{i}{2}\, dU_{I_m}(\widehat{D}_{ka}), \tag{1.8.1}$$

$$A^-_{lb} := dU_{I_m}(Y^-_{lb}) = \frac{1}{2}\, dU_{I_m}(D_{lb}) - \frac{i}{2}\, dU_{I_m}(\widehat{D}_{lb})) \tag{1.8.2}$$

and

$$C_{kl} := dU_{I_m}(Z^0_{kl}) = -i\, dU_{I_m}(D^0_{kl}). \tag{1.8.3}$$

By Lemma 2, we have

$$\begin{aligned} [A^+_{ka}, A^-_{lb}] &= \delta_{ab}\, C_{kl}, \\ [A^+_{ka}, A^+_{lb}] &= [A^-_{ka}, A^-_{lb}] = 0, \\ [C_{kl}, C_{mn}] &= [C_{kl}, A^+_{ma}] = [C_{kl}, A^-_{ma}] = 0. \end{aligned}$$

In particular, we have

$$[A^+_{ka}, A^-_{ka}] = 2\pi \cdot \mathrm{Id}, \qquad 1 \le k \le m, \quad 1 \le a \le n. \tag{1.8.4}$$

We note that A^+_{ka} and A^-_{lb} acts on the Schwartz space $C^\infty(\mathbb{R}^{(m,n)})$ or $\mathcal{S}(\mathbb{R}^{(m,n)})$ as the following linear differential operators

$$A^+_{ka} = \frac{1}{2}\left(\frac{\partial}{\partial \xi_{ka}} - 4\,\pi\, \xi_{ka}\right) \tag{1.8.5}$$

and

$$A_{lb}^{-} = \frac{1}{2}\left(\frac{\partial}{\partial \xi_{lb}} - 4\,\pi\,\xi_{lb}\right), \tag{1.8.6}$$

where $1 \leq k,l \leq m$ and $1 \leq a,b \leq n$. The differential operators A_{ka}^{+} and A_{lb}^{-} are called the creating operator of energy quantum and the annihilation operator of energy quantum respectively. It is easy to see that the adjoint of A_{ka}^{-} is $-A_{ka}^{+}$.

We start with the ground state $f_0(\xi)$ given by

$$f_0(\xi) = (\sqrt{2})^{mn} e^{-2\pi \sum_{k=1}^{m}\sum_{a=1}^{n} \xi_{ka}^2}. \tag{1.8.7}$$

By an easy computation, we have

$$\langle F_0, f_0\rangle = 1, \quad A_{ka}^{-}(f_0) = 0 \tag{1.8.8}$$

for all $1 \leq k \leq m$ and $1 \leq a \leq n$. This means that f_0 is a unit vecter in $L^2(\mathbb{R}^{(m,n)}, d\xi)$ which is annihilated by the annihilation operator A_{ka}^{-} : $\mathcal{S}(\mathbb{R}^{(m,n)}) \longrightarrow \mathcal{S}(\mathbb{R}^{(m,n)})$. For any $J \in \mathbb{Z}_{\geq 0}^{(m,n)}$, we define

$$f_J(\xi) := (A^{+})^{J} f_0(\xi) := (A_{11}^{+})^{J_{11}} \cdots (A_{ka}^{+})^{J_{ka}} \cdots (A_{mn}^{+})^{J_{mn}}\, f_0(\xi). \tag{1.8.9}$$

We give a lexicographic orderring on $\mathbb{Z}_{\geq 0}^{(m,n)}$. That is, for $J, K \in \mathbb{Z}_{\geq 0}^{(m,n)}$, $J < K$ if and only if $J_{11} = K_{11}, \cdots, J_{ij} = K_{ij}, J_{i,j+1} < K_{i,j+1}, \cdots$.

Lemma 21. *For each k, a with $1 \leq k \leq m$ and $1 \leq a \leq n$, we have*

$$A_{ka}^{-}(f_J) = -2\,\pi\, J_{ka}\, f_{J-\epsilon_{ka}}. \tag{1.8.10}$$

Proof. We prove this by induction on J. If $J = (0, \cdots, 0)$, (1.8.10) holds. Suppose (1.8.10) holds for J. For $\tilde{J} = J + \epsilon_{ka}$,

$$\begin{aligned} A_{ka}^{-}\left(f_{J+\epsilon_{ka}}\right) &= A_{ka}^{-} \circ A_{ka}^{+}(f_J) \\ &= \left(A_{ka}^{+} \circ A_{ka}^{-} - [A_{ka}^{+}, A_{ka}^{-}]\right)(f_J) \\ &= A_{ka}^{+}(-2\,\pi\, J_{ka}\, f_{J-\epsilon_{ka}}) - 2\,\pi\, f_J \\ &= -2\,\pi\, J_{ka}\, f_J - 2\,\pi\, f_J \\ &= -2\,\pi\,(J_{ka}+1)\, f_J. \end{aligned}$$

This completes the proof. □

Lemma 22.

$$\langle f_J, f_K\rangle = \begin{cases} (2\,\pi)^J\, J! & \textit{if } J = K \\ 0 & \textit{otherwise.} \end{cases}$$

Proof. If $J > K$, we have

$$\begin{aligned}\langle f_J, f_K \rangle &= \langle (A^+)^J f_0, (A^+)^K f_0 \rangle \\ &= (-1)^J \langle f_0, (A^-)^J \circ (A^+)^K f_0 \rangle \\ &= 0 \qquad \text{(by Lemma 21)}.\end{aligned}$$

In case $J < K$, $\langle f_J, f_K \rangle = \langle f_K, f_J \rangle = 0$. In case when $J = K$, we prove the above identity by induction on J. If $J = (0, 0, \cdots, 0)$, then $\langle f_0, f_0 \rangle = 1$. Assume that $(f_J, f_J) = (2\pi)^J J!$. Then according to (1.8.4) and Lemma 21, we have,

$$\begin{aligned}\langle f_{J+\epsilon_{ka}}, f_{J+\epsilon_{ka}} \rangle &= \langle A_{ka}^+(f_J), A_{ka}^+(f_J) \rangle \\ &= -\langle f_J, A_{ka}^- \circ A_{ka}^+(f_J) \rangle \\ &= -\langle f_J, (A_{ka}^+ \circ A_{ka}^- - [A_{ka}^+, A_{ka}^-]) f_J \rangle \\ &= -\langle f_J, -2\pi J_{ka} f_J - 2\pi f_J \rangle \\ &= 2\pi (J_{ka} + 1) \langle f_J, f_J \rangle \\ &= (2\pi)^{J+\epsilon_{ka}} (J + \epsilon_{ka})!.\end{aligned}$$

□

We define the normalized function $h_J \in \mathcal{S}(\mathbb{R}^{(m,n)})$ by

$$h_J := \left(\frac{1}{\sqrt{2\pi}} \right)^J (J!)^{-1/2} f_J, \quad J \in \mathbb{Z}_{\geq 0}^{(m,n)}. \tag{1.8.11}$$

Lemma 23. *For each $J \in \mathbb{Z}_{\geq 0}^{(m,n)}$ and all $k, a \in \mathbb{Z}$ with $1 \leq k \leq m$ and $1 \leq a \leq n$, we have*

$$A_{ka}^+(h_J) = \{2\pi (J_{ka} + 1)\}^{1/2} h_{J+\epsilon_{ka}} \tag{1.8.12}$$

and

$$A_{ka}^-(h_J) = -(2\pi J_{ka})^{1/2} h_{J-\epsilon_{ka}}. \tag{1.8.13}$$

Proof. According to (1.8.9), we have

$$\begin{aligned}A_{ka}^+(h_J) &= \left(\frac{1}{\sqrt{2\pi}} \right)^J (J!)^{-1/2} f_{J+\epsilon_{ka}} \\ &= (2\pi)^{1/2} (J_{ka} + 1)^{1/2} h_{J+\epsilon_{ka}} \\ &= \{2\pi (J_{ka} + 1)\}^{1/2} h_{J+\epsilon_{ka}}.\end{aligned}$$

According to Lemma 21, we have

$$
\begin{aligned}
A_{ka}^{-}(h_J) &= \left(\frac{1}{\sqrt{2\pi}}\right)^{J} (J!)^{-1/2} A_{ka}^{-}(f_J) \\
&= \left(\frac{1}{\sqrt{2\pi}}\right)^{J} (J!)^{-1/2} (-2\pi) J_{ka} f_{J-\epsilon_{ka}} \\
&= -(2\pi\, J_{ka})^{1/2} \left(\frac{1}{\sqrt{2\pi}}\right)^{J-\epsilon_{ka}} \{(J-\epsilon_{ka})!\}^{-1/2} f_{J-\epsilon_{ka}} \\
&= -(2\pi\, J_{ka})^{1/2} h_{J-\epsilon_{ka}}.
\end{aligned}
$$

□

Lemma 24. *For each $J \in \mathbb{Z}_{\geq 0}^{(m,n)}$ and $k, a \in \mathbb{Z}^{+}$ with $1 \leq k \leq m$ and $1 \leq a \leq n$, we have*

$$
A_{ka}^{+} \circ A_{ka}^{-}(h_J) = -2\pi\, J_{ka} h_J, \tag{1.8.14}
$$

$$
A_{ka}^{-} \circ A_{ka}^{+}(h_J) = -2\pi\,(J_{ka}+1)\, h_J. \tag{1.8.15}
$$

Proof. It follows immediately from (1.8.12) and (1.8.13).

$$
\begin{aligned}
A_{ka}^{+} \circ A_{ka}^{-}(h_J) &= -(2\pi\, J_{ka})^{1/2}\, A_{ka}^{+}(h_{J-\epsilon_{ka}}) \\
&= -(2\pi\, J_{ka})^{1/2}\, (2\pi J_{ka})^{1/2}\, h_J \\
&= -(2\pi\, J_{ka})\, h_J, \\
A_{ka}^{-} \circ A_{ka}^{+}(h_J) &= \{2\pi\,(J_{ka}+1)\}^{1/2}\, A_{ka}^{-}(h_{J+\epsilon_{ka}}) \\
&= \{2\pi\,(J_{ka}+1)\}^{1/2}\, (-1)\, \{2\pi(J_{ka}+1)\}^{1/2}\, h_J \\
&= -2\pi(J_{ka}+1)\, h_J.
\end{aligned}
$$

□

The linear differential operators

$$
A_{ka}^{+} \circ A_{ka}^{-} = \frac{1}{4}\left(\frac{\partial^2}{\partial \xi_{ka}^2} - 16\,\pi^2\,\xi_{ka}^2 + 4\,\pi\right)
$$

and

$$
A_{ka}^{-} \circ A_{ka}^{+} = \frac{1}{4}\left(\frac{\partial^2}{\partial \xi_{ka}^2} - 16\,\pi^2\,\xi_{ka}^2 - 4\,\pi\right)
$$

are called the **number operators** for the family $\{h_J \mid J \in \mathbb{Z}_{\geq 0}^{(m,n)}\}$. Now we consider the so-called *Hermite differential operator*

$$
H_{ka} := -2\left(A_{ka}^{+} \circ A_{ka}^{-} + A_{ka}^{-} \circ A_{ka}^{+}\right) = -\frac{\partial^2}{\partial \xi_{ka}^2} + 16\,\pi^2\,\xi_{ka}^2.
$$

H_{ka} is also called the **Schrödinger Hamiltonian** for the harmonic oscillator system in quantum mechanics. Obviously we have

$$H_{ka}(h_J) = 8\,\pi\,\left(J_{ka} + \frac{1}{2}\right)\,h_J, \qquad J \in \mathbb{Z}^{(m,n)}_{\geq 0}. \tag{1.8.16}$$

Thus the $\{\, h_J \,|\, J \in \mathbb{Z}^{(m,n)}_{\geq 0} \,\}$ is the set of normalized eigenforms of all Hermite operators H_{ka} with eigenvalues $\{8\,\pi\,(J_{ka} + \frac{1}{2}) \mid J \in \mathbb{Z}^{(m,n)}\}$. In other words, each h_J $(J \in \mathbb{Z}^{(m,n)})$ is the harmonic oscillator wave function with equidistant energies $\{8\,\pi\,(J_{ka} + \frac{1}{2}) \mid 1 \leq k \leq m,\ 1 \leq a \leq n\}$ in natural units. The Hermite operator H_{ka} acts on the Schwartz space $\mathcal{S}(\mathbb{R}^{(m,n)}) \subset L^2\big(\mathbb{R}^{(m,n)}, d\xi\big)$ and is self-adjoint.

Lemma 25. *For each $J \in \mathbb{Z}^{(m,n)}_{\geq 0}$ and $k, a \in \mathbb{Z}$ with $1 \leq k \leq m$ and $1 \leq a \leq n$,*

$$h_J(-\xi) = (-1)^J\,h_J(\xi), \tag{1.8.17}$$

$$\left(\frac{\partial}{\partial \xi_{ka}} - 4\,\pi\,\xi_{ka}\right) h_J(\xi) = 2\,\{2\,\pi\,(J_{ka}+1)\}^{1/2}\,h_{J+\epsilon_{ka}}(\xi), \tag{1.8.18}$$

$$\widehat{h_J} = (-i)^J\,h_J, \tag{1.8.19}$$

$$CF(h_J) = i^J\,h_J. \tag{1.8.20}$$

Thus $\widehat{h_J}$ and $CF(h_J)$ satisfy the differential equation (1.8.18). Here $\widehat{f}(\eta)$ denotes the Fourier transform of $f(\xi)$ on $\mathbb{R}^{(m,n)}$ defined by

$$\widehat{f}(\eta) := \int_{\mathbb{R}^{(m,n)}} f(\xi)\,e^{-2\,\pi\,i\,\langle \xi,\eta\rangle}\,d\xi, \quad \eta \in \mathbb{R}^{(m,n)}$$

and $CF(f)$ denotes the Fourier cotransform of f on $\mathbb{R}^{(m,n)}$ defined by

$$CF(f)(\xi) := \int_{\mathbb{R}^{(m,n)}} f(\eta)\,e^{2\,\pi\,i\,\langle \eta,\xi\rangle}\,d\eta, \quad \xi \in \mathbb{R}^{(m,n)}.$$

Proof. (1.8.17) is obvious. (1.8.18) follows immediately from (1.8.5) and (1.8.12). (1.8.19) and (1.8.20) follow from a simple computation. □

For $\xi = (\xi_{ka}) \in \mathbb{R}^{(m,n)}$, we briefly put $|\xi|^2 := \sum_{k=1}^m \sum_{a=1}^n \xi_{ka}^2$. We define the functions P_J $\left(J \in \mathbb{Z}^{(m,n)}_{\geq 0}\right)$ by

$$h_J(\xi) := P_J(\xi)\,e^{-2\,\pi\,|\xi|^2}, \qquad \xi \in \mathbb{R}^{(m,n)}. \tag{1.8.21}$$

Indeed, $P_J(\xi)$ are the Hermite polynomials of degree $J = (J_{11}, \cdots, J_{mn})$ normalized in such a way that they form an orthonormal family in $L^2\big(\mathbb{R}^{(m,n)}, e^{-4\pi|\xi|^2}d\xi\big)$ (it will proved later).

Lemma 26. *For each $J \in \mathbb{Z}^{(m,n)}_{\geq 0}$ and $k, a \in \mathbb{Z}^+$ with $1 \leq k \leq m$, and $1 \leq a \leq n$, we have*

$$\frac{\partial P_J(\xi)}{\partial \xi_{ka}} - 8\,\pi\,\xi_{ka}\,P_J(\xi) - 2\,\{2\,\pi\,(J_{ka}+1)\}^{1/2}\,P_{J+\epsilon_{ka}}(\xi) = 0 \qquad (1.8.22)$$

and

$$\frac{\partial P_{J+\epsilon_{ka}}(\xi)}{\partial \xi_{ka}} + 2\,\{2\,\pi\,(J_{ka}+1)\}^{1/2}\,P_J(\xi) = 0. \qquad (1.8.23)$$

Proof. (1.8.22) follows from (1.8.18). (1.8.23) follows from (1.8.6) and (1.8.13). □

Differentiating (1.8.22) with respect to ξ_{ka}, and then using (1.8.23), we see that $P_J(\xi)$ satisfies the so-called *Hermite equation*.

$$\frac{\partial^2 P_J(\xi)}{\partial \xi_{ka}^2} - 8\,\pi\,\xi_{ka}\,\frac{\partial P_J(\xi)}{\partial \xi_{ka}} + 8\,\pi\,J_{ka}\,P_J(\xi) = 0, \qquad (1.8.24)$$

where $J \in \mathbb{Z}_{\geq 0}^{(m,n)}$, $1 \leq k \leq m$ and $1 \leq a \leq n$. We set $\partial_{ka} := \frac{\partial}{\partial \xi_{ka}}$. Then (1.8.24) becomes

$$\partial_{ka}^2 P_J(\xi) - 8\,\pi\,\xi_{ka}\,\partial_{ka}\,P_J(\xi) + 8\pi J_{ka} P_J(\xi) = 0.$$

Differentiating (1.8.18) with respect to ξ_{ka}, we obtain

$$\begin{aligned} &\partial_{ka}^2 h_J(\xi) - 4\,\pi\,\xi_{ka}\,\partial_{ka} h_J(\xi) - 4\,\pi\,h_J(\xi) \\ &\qquad -2\,\{2\,\pi\,(J_{ka}+1)\}^{1/2}\,\partial_{ka} h_{J+\epsilon_{ka}}(\xi) = 0. \end{aligned} \qquad (1.8.25)$$

By the way, according to (1.8.23), we have

$$\begin{aligned} \partial_{ka} h_{J+\epsilon_{ka}}(\xi) &= \partial_{ka} P_{J+\epsilon_{ka}}(\xi)\,e^{-2\pi|\xi|^2} - 4\,\pi\,\xi_{ka}\,P_{J+\epsilon_{ka}}(\xi)\,e^{-2\pi|\xi|^2} \\ &= -\,2\,\{2\,\pi\,(J_{ka}+1)\}^{1/2}\,h_J(\xi) - 4\,\pi\,\xi_{ka}\,h_{J+\epsilon_{ka}}(\xi). \end{aligned}$$

If we substitute this relation into (1.8.25), we obtain

$$\partial_{ka}^2 h_J(\xi) - 16\,\pi^2\,\xi_{ka}^2\,h_J(\xi) = -8\,\pi\,\left(J_{ka} + \frac{1}{2}\right)\,h_J(\xi). \qquad (1.8.26)$$

Theorem 10. *The set* $\left\{ h_J \mid J \in \mathbb{Z}_{\geq 0}^{(m,n)} \right\}$ *of normalized Hermitian function in* $\mathcal{S}(\mathbb{R}^{(m,n)})$ *forms an orthonormal basis of* $L^2\big(\mathbb{R}^{(m,n)}, d\xi\big)$. *These* h_J *are eigenfunctions of the Fourier transform and the Fourier cotransform with eigenvalues* $(-i)^J$ *and* i^J *respectively.*

Proof. If X is a left-invariant vector field on G, we set, for brevity

$$U(X) := dU_{I_m}(X).$$

We will prove that the set

$$\left\{ U\left(\exp_G \left(\sum_{k,a} x_{ka} D_{ka} + \sum_{l,b} y_{lb} \hat{D}_{lb} \right)\right)(f_0) \,\middle|\, x_{ka}, y_{lb} \in \mathbb{R} \right\}$$

$(1 \le k, l \le m,\ 1 \le a, b \le n)$ is contained to the closed vector subspace of $L^2(\mathbb{R}^{(m,n)}, d\xi)$ spanned by the set $\{h_J \,|\, J \in \mathbb{Z}_{\ge 0}^{(m,n)}\}$ and the subspace generated by the above set is invariant under the action of U. Since the Schrödinger representation $(U_{I_m}, L^2(\mathbb{R}^{(m,n)}, d\xi))$ is irreducible, we conclude that the set $\{h_J \,|\, J \in \mathbb{Z}_{\ge 0}^{(m,n)}\}$ is a *complete* orthonormal basis for $L^2(\mathbb{R}^{(m,n)}, d\xi)$.

According to the commutation relation among D_{kl}^0, D_{ka}, $\widehat{D}_{lb}$ (cf. Lemma 2.1) and the fact that $U(D_{kl}^0)f = 2\pi i\, \delta_{kl}\, f$ for all $f \in \mathcal{S}(\mathbb{R}^{(m,n)})$, it suffices to prove the case $m = 1$ and $n = 1$. We put $D^0 := D_{11}^0$, $D := D_{11}$ and $\widehat{D} := \widehat{D}_{11}$. In other words, it remains to prove that the set

$$\left\{ U(\exp_G(xD + y\widehat{D}))(f_0) \,\middle|\, x, y \in \mathbb{R} \right\}$$

is contained in the closed vector subspace of $L^2(\mathbb{R}, d\xi)$ spanned by the set $\{h_j \,|\, j = 0, 1, 2, \cdots\}$.

First we note that by (1.8.1) and (1.8.2)

$$A^+ = \frac{1}{2}\left(U(D) + i\,U(\widehat{D})\right) \quad \text{and} \quad A^- = \frac{1}{2}\left(U(D) - i\,U(\widehat{D})\right).$$

For the present time being, we fix real numbers $x, y \in \mathbb{R}$. We put $z = x + iy \in \mathbb{C}$. It is obvious that $U(xD + y\widehat{D}) = \bar{z}A^+ + zA^-$. For all integers $k \ge 0$, $\ell \ge 0$ with $0 \le k \le \ell$, We define the complex numbers $c_{k\ell}$ by

$$U(xD + y\hat{D})^\ell (f_0) = \sum_{k=0}^{\ell} c_{k,\ell} f_k.$$

By the fact that $A^-(f_0) = 0$ and by (1.8.10), we have

$$\begin{aligned} U(xD + y\widehat{D})^{\ell+1}(f_0) &= (\bar{z}A^+ + zA^-)\left(\sum_{k=0}^{\ell} c_{k,\ell} f_k \right) \\ &= \sum_{k=0}^{\ell} c_{k,\ell}\,(\bar{z} f_{k+1} - 2\pi\, k\, z\, f_{k-1}). \end{aligned}$$

Thus we get the recurrence formula

$$c_{k,\ell+1} = \bar{z} c_{k-1,\ell} - 2\pi\,(k+1)\, z\, c_{k+1,\ell}, \qquad 1 \le k \le \ell - 1.$$

Let $z = |z|\, e^{2\pi i \varphi}$ with $\varphi \in [0, 1)$ for $z \ne 0$. We put

$$d_{k,\ell} := \left(|z|^{-1/2}\, e^{\pi\, i\, \varphi}\right)^{\ell+k} \left((2\,\pi\,|z|)^{-1/2}\, e^{-\pi\, i\,(\varphi-\frac{1}{2})}\right)^{\ell-k} c_{k,\ell}.$$

Then we have the recurrence formula

$$d_{k,\ell+1} = d_{k-1,\ell} + (k+1)\, d_{k+1,\ell}, \qquad 1 \le k \le \ell - 1.$$

For $1 \le k \le \ell - 1$, we put

$$b_{k,\ell} := d_{\ell-k,\ell}.$$

Then we get the recurrence formula

$$b_{k,\ell} = b_{k,\ell-1} + (\ell-k+1)\, b_{k-2,\ell-1}, \qquad 2 \le k \le \ell - 1.$$

If the starting value is $b_{0,0}$ and we define $b_{k,0} = 0$ for $k \ge 1$, then we get

$$b_{2p+1,\ell} = 0 \qquad \text{for } 0 \le p \le \frac{1}{2}(\ell-1)$$

and

$$b_{2p,\ell} = \frac{\ell!}{2^p\, p!\, (\ell-2p)!} \quad \text{for } 0 \le p \le \frac{1}{2}\ell.$$

So we obtain

$$\begin{aligned}
& U(xD + y\widehat{D})^{\ell}(f_0) = \sum_{k=0}^{\ell} c_{k,\ell} f_k \\
= & \sum_{k=0}^{\ell} \left(|z|^{1/2}\, e^{-\pi i \varphi}\right)^{\ell+k} \left((2\,\pi\,|z|)^{1/2}\, e^{\pi\, i\,(\varphi-\frac{1}{2})}\right)^{\ell-k} d_{k,\ell} f_k \\
= & \sum_{k=0}^{\ell} \left(|z|^{1/2}\, e^{-\pi i \varphi}\right)^{2\ell-k} \left((2\,\pi\,|z|)^{1/2}\, e^{\pi\, i\,(\varphi-\frac{1}{2})}\right)^{k} b_{k,\ell}\, f_{\ell-k} \\
= & \sum_{p=0}^{[\frac{\ell}{2}]} \left(|z|^{1/2}\, e^{-\pi\, i\, \varphi}\right)^{2\ell-2p} \left((2\,\pi\,|z|)^{1/2}\, e^{\pi\, i\,(\varphi-\frac{1}{2})}\right)^{2p} b_{2p,\ell}\, f_{\ell-2p} \\
= & \sum_{p=0}^{[\frac{\ell}{2}]} \bar{z}^{\ell-p} (-2\,\pi\, z)^p\, \frac{\ell!}{2^p\, p!\, (\ell-2p)!}\, f_{\ell-2p} \\
= & \sum_{p=0}^{[\frac{\ell}{2}]} \bar{z}^{\ell-p} (-\pi\, z)^p\, \frac{\ell!}{p!\, (\ell-2p)!}\, f_{\ell-2p}.
\end{aligned}$$

Thus we get

$$
\begin{aligned}
e^{U(xD+y\widehat{D})}(f_0) &= \sum_{\ell=0}^{\infty} \frac{1}{\ell!}\, U(xD+y\widehat{D})^{\ell}(f_0) \\
&= \sum_{\ell=0}^{\infty} \sum_{p=0}^{[\frac{\ell}{2}]} \frac{1}{\ell!}\, \bar{z}^{\ell-p}\, (-\pi z)^p\, \frac{\ell!}{p!\,(\ell-2p)!}\, f_{\ell-2p} \\
&= \sum_{\ell=0}^{\infty} \sum_{p=0}^{[\frac{\ell}{2}]} \frac{1}{p!\,(\ell-2p)!}\, (-\pi|z|^2)^p\, \bar{z}^{\ell-2p}\, f_{\ell-2p} \\
&= \sum_{k=0}^{\infty} \left\{ \sum_{p=0}^{\infty} \frac{1}{p!}\, (-\pi\,|z|^2)^p \right\} \frac{\bar{z}^k}{k!}\, f_k \\
&= e^{-\pi|z|^2} e^{\bar{z}A^+}(f_0) \\
&= e^{-\pi|z|^2} \sum_{k=0}^{\infty} \frac{(\sqrt{2\pi}\,\bar{z})^k}{(k!)^{1/2}}\, h_k.
\end{aligned}
$$

Therefore $U(\exp_G(xD+y\widehat{D}))(f_0)$ belongs to the closed subspace of $L^2(\mathbb{R}, d\xi)$ spanned by the set $\{\, h_j \mid j = 0, 1, 2, \cdots \}$. The latter part of the theorem follows immediately from (1.8.19) and (1.8.20). This completes the proof. □

Corollary 1. *The set* $\{\, P_J \mid J \in \mathbb{Z}_{\geq 0}^{(m,n)} \,\}$ *of Hermite polynomials forms an orthonormal basis for the* L^2*-space* $L^2\big(\mathbb{R}^{(m,n)}, e^{-4\pi|\xi|^2} d\xi\big)$.

Proof. The proof follows immediately from Theorem 10 and (1.8.21). □

1.9 Harmonic Analysis on the Quotient Space of $H_{\mathbb{R}}^{(n,m)}$

We fix an element $\Omega \in \mathbb{H}_n$ once and for all. Let $\mathcal{M}$ be a positive symmetric half-integral matrix of degree m. Let $L^2\big(\mathbb{R}^{(m,n)}, d\xi_{\Omega,\mathcal{M}}\big)$ be the L^2-space of $\mathbb{R}^{(m,n)}$ with respect to the measure

$$d\xi_{\Omega,\mathcal{M}} = e^{\pi\, i\, \sigma\{\mathcal{M}\xi(\Omega-\overline{\Omega})\,{}^t\xi\}}\, d\xi.$$

It is easy to show that the transformation $f(\xi) \longmapsto e^{\pi\, i\, \sigma\{\mathcal{M}\xi\Omega\,{}^t\xi\}} f(\xi)$ of $L^2\big(\mathbb{R}^{(m,n)}, d\xi_{\Omega,\mathcal{M}}\big)$ into $L^2\big(\mathbb{R}^{(m,n)}, d\xi\big)$ is an isomorphism. Since the set $\big\{\xi^J \mid J \in \mathbb{Z}_{\geq 0}^{(m,n)}\big\}$ is a basis of $L^2\big(\mathbb{R}^{(m,n)}, d\xi_{\Omega,\mathcal{M}}\big)$, the set

$$\Big\{\, e^{\pi\, i\, \sigma\{\mathcal{M}\xi\Omega\,{}^t\xi\}}\, \xi^J \mid J \in \mathbb{Z}_{\geq 0}^{(m,n)} \,\Big\}$$

is a basis of $L^2\big(\mathbb{R}^{(m,n)}, d\xi\big)$. We observe that there exists a canonical bijection A from the cosets $\mathcal{T} := \mathbb{Z}^{(m,n)}/(2\mathcal{M})\mathbb{Z}^{(m,n)}$. We denote by A_α the image of $\alpha \in \mathcal{T}$ under the bijection A.

For brevity, we put $G = H_{\mathbb{R}}^{(n,m)}$. For each $A_\alpha \in \mathcal{L}$ and each $J \in \mathbb{Z}_{\geq 0}^{(m,n)}$, we define a function $\Phi_J^{(\mathcal{M})} \begin{bmatrix} A_\alpha \\ 0 \end{bmatrix} (\Omega| \cdot)$ on G by

$$\begin{aligned} &\Phi_J^{(\mathcal{M})} \begin{bmatrix} A_\alpha \\ 0 \end{bmatrix} \big(\Omega|(\lambda,\mu,\kappa)\big) : \\ &= e^{2\pi i\, \sigma\{\mathcal{M}(\kappa-\lambda\,{}^t\mu)\}} \sum_{N\in\mathbb{Z}^{(m,n)}} (\lambda+N+A_\alpha)^J \\ &\quad \times\, e^{2\pi i\, \sigma\{\mathcal{M}((\lambda+N+A_\alpha)\Omega\,{}^t(\lambda+N+A_\alpha)\,+\,2(\lambda+N+A_\alpha)\,{}^t\mu)\}}, \end{aligned} \tag{1.9.1}$$

where $(\lambda,\mu,\kappa) \in G$. We let $\Gamma_G = H_{\mathbb{Z}}^{(n,m)}$ be the discrete subgroup of G consisting of integral elements. That is,

$$\Gamma_G = \{(\lambda,\mu,\kappa) \in G \mid \lambda,\mu,\kappa \text{ integral}\}.$$

According to [42], Proposition 1.3, the function $\Phi_J^{(\mathcal{M})} \begin{bmatrix} A_\alpha \\ 0 \end{bmatrix} \big(\Omega|(\lambda,\mu,\kappa)\big)$ satisfies the transformation behaviour

$$\Phi_J^{(\mathcal{M})} \begin{bmatrix} A_\alpha \\ 0 \end{bmatrix} \big(\Omega|\gamma\circ g\big) = \Phi_J^{(\mathcal{M})} \begin{bmatrix} A_\alpha \\ 0 \end{bmatrix} (\Omega|g) \tag{1.9.2}$$

holds for all $\gamma \in \Gamma_G$ and $g \in G$. Thus the functions

$$\Phi_J^{(\mathcal{M})} \begin{bmatrix} A_\alpha \\ 0 \end{bmatrix} (\Omega|(\lambda,\mu,\kappa)) \; \Big(J \in \mathbb{Z}_{\geq 0}^{(m,n)}\Big)$$

are real analytic functions on the quotient space $\Gamma_G\backslash G$. Let $H_\Omega^{(\mathcal{M})}\begin{bmatrix} A_\alpha \\ 0 \end{bmatrix}$ be the completion of the vector space spanned by

$$\Phi_J^{(\mathcal{M})}\begin{bmatrix} A_\alpha \\ 0 \end{bmatrix}(\Omega|(\lambda,\mu,\kappa)) \ \left(J \in \mathbb{Z}_{\geq 0}^{(m,n)}\right)$$

and let $\overline{H_\Omega^{(\mathcal{M})}\begin{bmatrix} A_\alpha \\ 0 \end{bmatrix}}$ be the complex conjugate of $H_\Omega^{(\mathcal{M})}\begin{bmatrix} A_\alpha \\ 0 \end{bmatrix}$.

Let $L^2(\Gamma_G\backslash G)$ be the L^2-space of the quotient space $\Gamma_G\backslash G$ with respect to the invariant measure

$$d\lambda_{11}\cdots d\lambda_{m,n-1}d\lambda_{mn}d\mu_{11}\cdots d\mu_{m,n-1}d\mu_{mn}d\kappa_{11}d\kappa_{12}\cdots d\kappa_{mm}.$$

Let ρ be he right regular representation of G on the Hilbert space $L^2(\Gamma_G\backslash G)$ given by

$$(\rho(g_0)\phi)(g) := \phi(gg_0), \ \ g_0, g \in G, \ \ \phi \in L^2(\Gamma_G\backslash G).$$

In [42], the author proved that the subspaces $H_\Omega^{(\mathcal{M})}\begin{bmatrix} A_\alpha \\ 0 \end{bmatrix}$ and $\overline{H_\Omega^{(\mathcal{M})}\begin{bmatrix} A_\alpha \\ 0 \end{bmatrix}}$ are irreducible invariant subspaces of $L^2(\Gamma_G\backslash G)$ with respect to ρ and the decomposition of the right regular representation ρ is given by

$$\begin{aligned} L^2(\Gamma_G\backslash G) = \bigoplus_{\mathcal{M},\alpha} H_\Omega^{(\mathcal{M})}\begin{bmatrix} A_\alpha \\ 0 \end{bmatrix} \oplus \left(\overline{\bigoplus_{\mathcal{M},\alpha} H_\Omega^{(\mathcal{M})}\begin{bmatrix} A_\alpha \\ 0 \end{bmatrix}} \right) \\ \oplus \left(\bigoplus_c R(c) \right) \oplus \left(\bigoplus_{k,\ell \in \mathbb{Z}^{(m,n)}} \mathbb{C}\, e^{2\pi i\, \sigma(k\,{}^t\lambda + \ell\,{}^t\mu)} \right), \end{aligned}$$

where $\mathcal{M}$ (respectively c) runs over the set of all positive symmetric half integral matrices of degree m (respectively the set of all half integral nonzero matrices of degree m which are neither positive nor negative definite), $R(c)$ is the sum of irreducible representations π_c which occur in ρ and A_α runs over a complete system of representatives of the cosets $(2\mathcal{M})^{-1}\mathbb{Z}^{(m,n)}/\mathbb{Z}^{(m,n)}$.

Lemma 27. *The transform of* $L^2\left(\mathbb{R}^{(m,n)}, d\xi_{\Omega,\mathcal{M}}\right)$ *onto* $H_\Omega^{(\mathcal{M})}\begin{bmatrix} A_\alpha \\ 0 \end{bmatrix}$ *given by*

$$\xi^J \longmapsto \Phi_J^{(\mathcal{M})}\begin{bmatrix} A_\alpha \\ 0 \end{bmatrix}(\Omega|(\lambda,\mu,\kappa)) \tag{1.9.3}$$

is an isomorphism of Hilbert spaces.

Proof. For the proof, we refer to [42], Lemma 3.2. □

We write

$$f_{\Omega,J}^{(\mathcal{M})}(\xi) := e^{2\pi i\, \sigma(\mathcal{M}\xi\Omega\,{}^t\xi)}\xi^J, \ \ J \in \mathbb{Z}_{\geq 0}^{(m,n)}. \tag{1.9.4}$$

We let $\Phi_{\Omega,\alpha}^{(\mathcal{M})}$ be the transform of $L^2\big(\mathbb{R}^{(m,n)}, d\xi\big)$ onto $H_\Omega^{(\mathcal{M})}\begin{bmatrix} A_\alpha \\ 0 \end{bmatrix}$ defined by

$$\Phi_{\Omega,\alpha}^{(\mathcal{M})}\left(f_{\Omega,J}^{(\mathcal{M})}\right) := \Phi_J^{(\mathcal{M})}\begin{bmatrix} A_\alpha \\ 0 \end{bmatrix}\big(\Omega|(\lambda,\mu,\kappa)\big). \tag{1.9.5}$$

Then $\Phi_{\Omega,\alpha}^{(\mathcal{M})}$ is an isometry of $L^2\big(\mathbb{R}^{(m,n)}, d\xi\big)$ onto $H_\Omega^{(\mathcal{M})}\begin{bmatrix} A_\alpha \\ 0 \end{bmatrix}$ such that

$$U^{S,\mathcal{M}}\left((\lambda,\mu,\kappa)\right)\circ \Phi_{\Omega,\alpha}^{(\mathcal{M})} = \Phi_{\Omega,\alpha}^{(\mathcal{M})}\circ U^{S,\mathcal{M}}\big((\lambda,-\mu,-\kappa)\big),$$

where $U^{S,\mathcal{M}}$ is the Schrödinger representation of G defined by (6.45).

Let $\Delta_{\Omega,\mathcal{M}}$ be the isometry of $L^2\big(\mathbb{R}^{(m,n)}, d\xi_{\Omega,\mathcal{M}}\big)$ onto $L^2\big(\mathbb{R}^{(m,n)}, d\xi\big)$ defined by

$$(\Delta_{\Omega,\mathcal{M}} f)(\xi) := e^{\pi i\,\sigma\{\mathcal{M}\xi\Omega\,{}^t\xi\}}\, f(\xi). \tag{1.9.6}$$

We define the unitary representation $U_\Omega^{S,\mathcal{M}}$ of G on $L^2\big(\mathbb{R}^{(m,n)}, d\xi_{\Omega,\mathcal{M}}\big)$ by

$$\left(U_\Omega^{S,\mathcal{M}}(g)f\right)(\xi) := \Delta_{\Omega,\mathcal{M}}^{-1}\left(\left(U_\Omega^{S,\mathcal{M}}(g)(\Delta_{\Omega,\mathcal{M}}f)\right)(\xi)\right), \tag{1.9.7}$$

where $f\in L^2\big(\mathbb{R}^{(m,n)}, d\xi_{\Omega,\mathcal{M}}\big)$ and $\xi\in\mathbb{R}^{(m,n)}$.

Now we write down the image of $f_{\Omega,J}^{(\mathcal{M})}\in L^2\big(\mathbb{R}^{(m,n)}, d\xi\big)$ under $\vartheta_{\mathcal{M},\alpha}$ (cf. (1.6.22)) explicitly.

$$\begin{aligned}
&\left(\vartheta_{\mathcal{M},\alpha} f_{\Omega,J}^{(\mathcal{M})}\right)((\lambda,\mu,\kappa)) \\
&= \left(\vartheta_{\mathcal{M},\alpha} f_{\Omega,J}^{(\mathcal{M})}\right)([\lambda,\mu,\kappa+\mu\,{}^t\lambda]) \\
&= e^{2\pi i\,\sigma\{\mathcal{M}(\kappa+\mu\,{}^t\lambda)\}} \sum_{N\in\mathbb{Z}^{(m,n)}} e^{2\pi i\,\sigma\{\mathcal{M}((\lambda+N)\Omega\,{}^t(\lambda+N)\,+\,2\,N\,{}^t\mu)\}}\,(\lambda+N)^J \\
&= e^{2\pi i\,\sigma\{\mathcal{M}(\kappa-\lambda\,{}^t\mu)\,+\,\alpha\,{}^t\mu\}} \\
&\quad\times \sum_{N\in\mathbb{Z}^{(m,n)}} e^{2\pi i\,\sigma\{\mathcal{M}((\lambda+N)\Omega\,{}^t(\lambda+N)\,+\,2\,(\lambda+N)\,{}^t\mu)\}}\,(\lambda+N)^J.
\end{aligned}$$

In particular, if $\alpha=0,\ \kappa=0$ and $J=0$, then we have

$$\begin{aligned}
&\left(\vartheta_{\mathcal{M},0} f_{\Omega,0}^{(\mathcal{M})}\right)((\lambda,\mu,0)) \\
&= e^{-2\pi i\,\sigma(\mathcal{M}\lambda\,{}^t\mu)} \sum_{N\in\mathbb{Z}^{(m,n)}} e^{2\pi i\,\sigma\{\mathcal{M}((\lambda+N)\Omega\,{}^t(\lambda+N)+2\,(\lambda+N)\,{}^t\mu)\}} \\
&= e^{2\pi\, i\,\sigma\{\mathcal{M}(\lambda\,\Omega\,{}^t\lambda+\lambda\,{}^t\mu)\}} \sum_{N\in\mathbb{Z}^{(m,n)}} e^{2\pi i\,\sigma\{\mathcal{M}(N\Omega\,{}^tN+2\,(\lambda\Omega+\mu){}^tN)\}} \\
&= e^{2\pi i\,\sigma\{\mathcal{M}(\lambda\,\Omega\,{}^t\lambda+\lambda\,{}^t\mu)\}}\,\vartheta^{(2\mathcal{M})}\begin{bmatrix}0\\0\end{bmatrix}(\Omega,\lambda\Omega+\mu).
\end{aligned}$$

Therefore we obtain

Proposition 7. *Let $\mathcal{M}$ be a positive symmetric half-integral matrix of degree m. Let $\alpha\in\mathcal{T}$ and $J\in\mathbb{Z}_{\geq 0}^{(m,n)}$. Then we have*

$$\begin{aligned}
&\left(\vartheta_{\mathcal{M},\alpha} f_{\Omega,J}^{(\mathcal{M})}\right)((\lambda,\mu,\kappa)) \\
&= e^{2\pi i\,\sigma\{(\kappa-\lambda\,{}^t\mu)+\alpha\,{}^t\mu\}} \sum_{N\in\mathbb{Z}^{(m,n)}} e^{2\pi i\,\sigma\{\mathcal{M}((\lambda+N)\Omega\,{}^t(\lambda+N)+2(\lambda+N)\,{}^t\mu)\}}(\lambda+N)^J.
\end{aligned}$$

In particular,

$$\left(\vartheta_{\mathcal{M},0} f_{\Omega,0}^{(\mathcal{M})}\right)((\lambda,\mu,0)) = e^{2\pi i\,\sigma\{\mathcal{M}(\lambda\Omega\,{}^t\lambda+\lambda\,{}^t\mu)\}}\,\vartheta^{(2\mathcal{M})}\begin{bmatrix}0\\0\end{bmatrix}(\Omega,\lambda\,\Omega+\mu).$$

It is easy to see that the following diagrams are commutative.

$$\begin{array}{ccc}
L^2\left(\mathbb{R}^{(m,n)},d\xi\right) & \xrightarrow{U^{S,\mathcal{M}}(g)} & L^2\left(\mathbb{R}^{(m,n)},d\xi\right) \\
\vartheta_{\mathcal{M},\alpha}\downarrow & & \downarrow\vartheta_{\mathcal{M},\alpha} \\
\mathcal{H}_{\mathcal{M},\alpha} & & \mathcal{H}_{\mathcal{M},\alpha} \\
\| & \xrightarrow{\pi_{\mathcal{M},\alpha}(g)} & \| \\
\pi_{\mathcal{M},\alpha} & & \pi_{\mathcal{M},\alpha}
\end{array}$$

diagram 10.1

$$\begin{array}{ccc}
L^2\left(\mathbb{R}^{(m,n)},d\xi_{\Omega,\mathcal{M}}\right) & \xrightarrow{U_{\Omega}^{S,\mathcal{M}}(g)} & L^2\left(\mathbb{R}^{(m,n)},d\xi_{\Omega,\mathcal{M}}\right) \\
\Delta_{\Omega,\mathcal{M}}\downarrow & & \downarrow\Delta_{\Omega,\mathcal{M}} \\
L^2\left(\mathbb{R}^{(m,n)},d\xi\right) & \xrightarrow{U^{S,\mathcal{M}}(g)} & L^2\left(\mathbb{R}^{(m,n)},d\xi\right)
\end{array}$$

diagram 10.2

$$\begin{array}{ccc} L^2\big(\mathbb{R}^{(m,n)}, d\xi\big) & \xrightarrow{U^{S,\mathcal{M}}(g)} & L^2\big(\mathbb{R}^{(m,n)}, d\xi\big) \\ {\scriptstyle I_{\mathcal{M}}}\downarrow & & \downarrow{\scriptstyle I_{\mathcal{M}}} \\ \mathcal{H}_{F,\mathcal{M}} & \xrightarrow{U^{F,\mathcal{M}}(g)} & \mathcal{H}_{F,\mathcal{M}} \end{array}$$

diagram 10.3

$$\begin{array}{ccc} L^2\big(\mathbb{R}^{(m,n)}, d\xi\big) & \xrightarrow{U^{S,\mathcal{M}}(g)} & L^2\big(\mathbb{R}^{(m,n)}, d\xi\big) \\ {\scriptstyle \Phi^{(\mathcal{M})}_{\Omega,\alpha}}\downarrow & & \downarrow{\scriptstyle \Phi^{(\mathcal{M})}_{\Omega,\alpha}} \\ H^{(\mathcal{M})}_{\Omega}\begin{bmatrix} A_\alpha \\ 0 \end{bmatrix} & \xrightarrow{\rho_{\mathcal{M}}(g)} & H^{(\mathcal{M})}_{\Omega}\begin{bmatrix} A_\alpha \\ 0 \end{bmatrix} \end{array}$$

diagram 10.4

Here $g \in G$ and $\rho_{\mathcal{M}}$ denotes the restriction of the right regular representation ρ to $H^{(\mathcal{M})}_{\Omega}\begin{bmatrix} A_\alpha \\ 0 \end{bmatrix}$. We know that the mapping $\vartheta_{\mathcal{M},\alpha}$, $\Delta_{\Omega,\mathcal{M}}$, $I_{\mathcal{M}}$ and $\Phi^{(\mathcal{M})}_{\Omega,\alpha}$ are all the isomorphisms preserving the norms. Hence we have

Theorem 11. *For each $\alpha \in \mathcal{T}$, $\Omega \in \mathbb{H}_n$ and $\mathcal{M}$ positive symmetric half-integral matrix of degree m, the following five representations are unitarily equivalent to each other via the intertwining operators $\vartheta_{\mathcal{M},\alpha}$, $I_{\mathcal{M}}$, $\Phi^{(\mathcal{M})}_{\Omega,\alpha}$ and $\Delta_{\Omega,\mathcal{M}}$:*

(1) the Schrödinger representation $\big(U^{S,\mathcal{M}}, L^2\big(\mathbb{R}^{(m,n)}, d\xi\big)\big)$,

(2) the lattice representation $(\pi_{\mathcal{M},\alpha}, \mathcal{H}_{\mathcal{M},\alpha})$,

(3) the Fock representation $\big(U^{F,\mathcal{M}}, \mathcal{H}_{F,\mathcal{M}}\big)$,

(4) the representation $\left(\rho_{\mathcal{M}}, H^{(\mathcal{M})}_{\Omega}\begin{bmatrix} A_\alpha \\ 0 \end{bmatrix}\right)$

and

(5) the representation $\big(U^{S,\mathcal{M}}_{\Omega}, L^2\big(\mathbb{R}^{(m,n)}, d\xi_{\Omega,\mathcal{M}}\big)\big)$.

Remark 5. The multiplicity of the Schrödinger representation $U^{S,\mathcal{M}}$ of G in $\big(\rho, L^2(\Gamma_G\backslash G)\big)$ is $(\det 2\mathcal{M})^n$.

We refer to [42] for detail. Theorem 11 may be pictured as follows.

$$\begin{array}{ccccc} & & \mathcal{M}_{F,\mathcal{M}} & & \\ & & \uparrow I_{\mathcal{M}} & & \\ L^2\big(\mathbb{R}^{(m,n)}, d\xi_{\Omega,\mathcal{M}}\big) & \xrightarrow{\Delta_{\mathcal{M},\alpha}} & L^2\big(\mathbb{R}^{(m,n)}, d\xi\big) & \xrightarrow{\vartheta_{\mathcal{M},\alpha}} & \mathcal{H}_{\mathcal{M},\alpha} \\ & & \downarrow \Phi_{\Omega,\alpha}^{(\mathcal{M})} & & \\ & & H_{\Omega}^{(\mathcal{M})}\begin{bmatrix} A_\alpha \\ 0 \end{bmatrix} & & \end{array}$$

figure 10.5

Finally we describe explicitly the orthonormal bases of

$$L^2\big(\mathbb{R}^{(m,n)}, d\xi\big),\ L^2\big(\mathbb{R}^{(m,n)}, d\xi_{\Omega,\mathcal{M}}\big),\ \mathcal{H}_{\mathcal{M},\alpha},\ \mathcal{H}_{F,\mathcal{M}} \text{ and } H_{\Omega}^{(\mathcal{M})}\begin{bmatrix} A_\alpha \\ 0 \end{bmatrix}$$

respectively.

In the previous section, we proved that the family of the functions

$$h_J(\xi) = \left(\frac{1}{\sqrt{2\pi}}\right)^J (J!)^{-1/2} f_J(\xi), \quad J \in \mathbb{Z}_{\geq 0}^{(m,n)}$$

forms an orthonormal basis of $L^2\big(\mathbb{R}^{(m,n)}, d\xi\big)$. Therefore the set

$$\left\{ e^{-\pi\, i\, \sigma(\mathcal{M}\xi\Omega\, {}^t\xi)}\, h_J(\xi) \mid J \in \mathbb{Z}_{\geq 0}^{(m,n)} \right\}$$

forms an orthonormal basis for $L^2\big(\mathbb{R}^{(m,n)}, d\xi_{\Omega,\mathcal{M}}\big)$. For each $J \in \mathbb{Z}_{\geq 0}^{(m,n)}$, the set of functions

$$\begin{aligned} &\vartheta_{\mathcal{M},\alpha,J}(\lambda,\mu,\kappa) : \\ &= (\vartheta_{\mathcal{M},\alpha} h_J)(\lambda,\mu,\kappa) \\ &= e^{2\pi i\, \sigma\{\mathcal{M}(\kappa + \mu\, {}^t\lambda + \alpha\, {}^t\mu)\}} \sum_{N \in \mathbb{Z}^{(m,n)}} h_J(\lambda + N)\, e^{4\pi i\, \sigma(\mathcal{M}N\, {}^t\mu)}, \quad J \in \mathbb{Z}_{\geq 0}^{(m,n)} \end{aligned}$$

forms an orthonormal basis for $\mathcal{H}_{\mathcal{M},\alpha}$. For each $J \in \mathbb{Z}_{\geq 0}^{(m,n)}$, we define the function

$$
\begin{aligned}
H_J^{(\mathcal{M})}\begin{bmatrix} A_\alpha \\ 0 \end{bmatrix}(\Omega|(\lambda,\mu,\kappa)) &:= 2^{\frac{mn}{4}-\frac{|J|}{2}}\,(J!)^{-1/2}\,(\det 2\mathcal{M})^{n/4}(\det \operatorname{Im}\Omega)^{m/4} \\
&\times e^{\pi i\,\sigma(\mathcal{M}(\kappa-\lambda\,{}^t\mu))} \sum_{N\in\mathbb{Z}^{(m,n)}} H_J\left(\sqrt{2\pi}\,(2\mathcal{M})^{1/2}\,(\lambda+N+A_\alpha)(\operatorname{Im}\Omega)^{1/2}\right) \\
&\times e^{\pi i\,\sigma\{\mathcal{M}((\lambda+N+A_\alpha)\Omega\,{}^t(\lambda+N+A_\alpha)+2\,(\lambda+N+A_\alpha)\,{}^t\mu)\}},
\end{aligned}
$$

where $H_J(\xi)$ is the Hermite polynomial on $\mathbb{R}^{(m,n)}$ in several variables defined by

$$H_J(\xi) := H_{J_{11}}(\xi_{11})H_{J_{12}}(\xi_{12})\cdots H_{J_{mn}}(\xi_{mn}).$$

It was proved in [43] that the functions $H_J^{(\mathcal{M})}\begin{bmatrix} A_\alpha \\ 0 \end{bmatrix}(\Omega|(\lambda,\mu,\kappa))$ $\left(J\in\mathbb{Z}_{\geq 0}^{(m,n)}\right)$ form an orthonormal basis for $H_\Omega^{(\mathcal{M})}\begin{bmatrix} A_\alpha \\ 0 \end{bmatrix}$.

Theorem 12. *We have an orthonormal basis for each of the above five representation:*

(1) The set $\left\{ h_J \mid J\in\mathbb{Z}_{\geq 0}^{(m,n)} \right\}$ *forms an orthonormal basis for* $L^2\left(\mathbb{R}^{(m,n)}, d\xi\right)$.

(2) The set $\left\{ e^{-\pi i\,\sigma(\mathcal{M}\xi\Omega\,{}^t\xi)}\, h_J \mid J\in\mathbb{Z}_{\geq 0}^{(m,n)} \right\}$ *forms an orthonormal basis for* $L^2\left(\mathbb{R}^{(m,n)}, d\xi_{\Omega,\mathcal{M}}\right)$.

(3) The set $\left\{ \vartheta_{\mathcal{M},\alpha,J} \mid J\in\mathbb{Z}_{\geq 0}^{(m,n)} \right\}$ *forms an orthonormal basis for* $\mathcal{H}_{\mathcal{M},\alpha}$.

(4) The set $\left\{ \Phi_{\mathcal{M},J} \mid J\in\mathbb{Z}_{\geq 0}^{(m,n)} \right\}$ *(cf. (1.5.36)) forms an orthonormal basis for* $\mathcal{H}_{F,\mathcal{M}}$.

(5) The set $H_J^{(\mathcal{M})}\begin{bmatrix} A_\alpha \\ 0 \end{bmatrix}(\Omega|(\lambda,\mu,\kappa))$ $\left(J\in\mathbb{Z}_{\geq 0}^{(m,n)}\right)$ *forms an orthonormal basis for* $H_\Omega^{(\mathcal{M})}\begin{bmatrix} A_\alpha \\ 0 \end{bmatrix}$.

Chapter 2
Theta Functions and the Weil Representation

2.1 The Symplectic Group

We recall that

$$Sp(n,\mathbb{R}) = \{M \in \mathbb{R}^{(2n,2n)} \mid {}^tMJ_nM = J_n\}$$

is the symplectic group of degree n, where

$$J_n = \begin{pmatrix} 0 & I_n \\ -I_n & 0 \end{pmatrix}.$$

If $M = \begin{pmatrix} A & B \\ C & D \end{pmatrix} \in Sp(n,\mathbb{R})$ with $A, B, C, D \in \mathbb{R}^{(n,n)}$, then

$${}^tAD - {}^tCB = I_n, \qquad {}^tAC = {}^tCA, \qquad {}^tBD = {}^tDB. \tag{2.1.1}$$

We note that $Sp(1,\mathbb{R}) = SL(2,\mathbb{R})$. The inverse of $M = \begin{pmatrix} A & B \\ C & D \end{pmatrix} \in Sp(n,\mathbb{R})$ is

$$M^{-1} = J_n^{-1}\,{}^tM\,J_n = \begin{pmatrix} {}^tD & -{}^tB \\ -{}^tC & {}^tA \end{pmatrix}.$$

Since $J_n^{-1} = -J_n$ and ${}^tM^{-1}J_nM^{-1} = J_n$ with $M \in Sp(n,\mathbb{R})$, we see that

$${}^tM^{-1}J_n^{-1}M^{-1} = J_n^{-1}, \quad \text{that is,} \quad MJ_n\,{}^tM = J_n.$$

Thus if $M \in Sp(n,\mathbb{R})$, then ${}^tM \in Sp(n,\mathbb{R})$. If $M = \begin{pmatrix} A & B \\ C & D \end{pmatrix} \in Sp(n,\mathbb{R})$, then

$$A\,{}^tD - B\,{}^tC = I_n, \qquad A\,{}^tB = B\,{}^tA, \qquad C\,{}^tD = D\,{}^tC. \tag{2.1.2}$$

Lemma 28. *Let* $M = \begin{pmatrix} A & B \\ C & D \end{pmatrix} \in Sp(n, \mathbb{R})$ *and* $\Omega \in \mathbb{H}_n$. *Then*

(a) $C\Omega + D$ *is nonsingular.*

(b) $(A\Omega + B)(C\Omega + D)^{-1}$ *is an element of* $\mathbb{H}_n$.

Proof. Let $\Omega = X + iY \in \mathbb{H}_n$ with $X, Y \in \mathbb{R}^{(n,n)}$ and $Y > 0$. Then

$$^t\begin{pmatrix} \Omega \\ I_n \end{pmatrix} J_n \begin{pmatrix} \Omega \\ I_n \end{pmatrix} = 0 \tag{2.1.3}$$

and

$$^t\begin{pmatrix} \Omega \\ I_n \end{pmatrix} J_n \overline{\begin{pmatrix} \Omega \\ I_n \end{pmatrix}} = 2\,iY > 0. \tag{2.1.4}$$

We set

$$S = A\Omega + B \qquad \text{and} \qquad T = C\Omega + D.$$

By (2.1.3), we have

$$\begin{aligned} {}^t\begin{pmatrix} S \\ T \end{pmatrix} J_n \begin{pmatrix} S \\ T \end{pmatrix} &= {}^t\left\{ M \begin{pmatrix} \Omega \\ I_n \end{pmatrix} \right\} J_n \left\{ M \begin{pmatrix} \Omega \\ I_n \end{pmatrix} \right\} \\ &= {}^t\begin{pmatrix} \Omega \\ I_n \end{pmatrix} {}^tMJ_nM \begin{pmatrix} \Omega \\ I_n \end{pmatrix} \\ &= {}^t\begin{pmatrix} \Omega \\ I_n \end{pmatrix} J_n \begin{pmatrix} \Omega \\ I_n \end{pmatrix} = 0. \end{aligned}$$

By (2.1.4), we have

$$\frac{1}{2i}\, {}^t\begin{pmatrix} S \\ T \end{pmatrix} J_n \overline{\begin{pmatrix} S \\ T \end{pmatrix}} = \frac{1}{2i}\, {}^t\begin{pmatrix} \Omega \\ I_n \end{pmatrix} J_n \overline{\begin{pmatrix} \Omega \\ I_n \end{pmatrix}} = Y > 0.$$

Thus we have

$$^tST - {}^tTS = 0 \qquad \text{and} \qquad \frac{1}{2i}\left({}^tS\overline{T} - {}^tT\overline{S} \right) = Y > 0. \tag{2.1.5}$$

Assume $Tv = (C\Omega + D)v = 0$ for some $v \in \mathbb{C}^n$. Then $\overline{T}\overline{v} = 0$, ${}^tv\,{}^tT = 0$ and hence

$$\frac{1}{2i}\, {}^tv\left({}^tS\overline{T} - {}^tT\overline{S} \right)\overline{v} = 0.$$

By (2.1.5), $v = 0$ and so $T = C\Omega + D$ is nonsingular. This proves the statement (a).

We set

$$\Omega_* = (A\Omega + B)(C\Omega + D)^{-1} = ST^{-1}.$$

By (2.1.5), we have $\Omega_* = {}^t\Omega_*$ and

$$\begin{aligned}\operatorname{Im}\Omega_* &= \frac{1}{2i}\big(\Omega_* - \overline{\Omega}_*\big) = \frac{1}{2i}\big({}^t\Omega_* - \overline{\Omega}_*\big)\\ &= \frac{1}{2i}\,{}^tT^{-1}\big({}^tS\overline{T} - {}^tT\overline{S}\big)\,\overline{T}^{-1}\\ &= {}^tT^{-1}Y\,\overline{T}^{-1} > 0.\end{aligned}$$

Therefore $\Omega_* \in \mathbb{H}_n$. This completes the proof of the statement (b). □

Lemma 29. *The symplectic group $Sp(n,\mathbb{R})$ acts on the Siegel upper half plane $\mathbb{H}_n$ transitively by*

$$M\cdot\Omega = (A\Omega + B)(C\Omega + D)^{-1}, \tag{2.1.6}$$

where $M = \begin{pmatrix} A & B\\ C & D\end{pmatrix} \in Sp(n,\mathbb{R})$ and $\Omega \in \mathbb{H}_n$.

Proof. Let $\Omega = X + i\,Y \in \mathbb{H}_n$ with $X, Y \in \operatorname{Sym}(n,\mathbb{R})$ and $Y > 0$. It suffices to show that there exists an element $M \in Sp(n,\mathbb{R})$ such that $M\cdot(iI_n) = \Omega$. We choose $Q \in GL(n,\mathbb{R})$ such that $Q^2 = Y$. We take

$$M = \begin{pmatrix} I_n & X\\ 0 & I_n\end{pmatrix}\begin{pmatrix} {}^tQ & 0\\ 0 & Q^{-1}\end{pmatrix}.$$

According to (2.1.2), $M \in Sp(n,\mathbb{R})$. Clearly $M\cdot(iI_n) = X + i\,Y = \Omega$. □

It is known (cf. [7], p. 322-328, [15], p. 10) that $Sp(n,\mathbb{R})$ is generated by the following elements

$$\begin{aligned} t_b &= \begin{pmatrix} I_n & b\\ 0 & I_n\end{pmatrix} \quad \text{with } b = {}^tb \in \mathbb{R}^{(n,n)},\\ d_a &= \begin{pmatrix} {}^ta & 0\\ 0 & a^{-1}\end{pmatrix} \quad \text{with } a \in GL(n,\mathbb{R}),\\ \sigma_n &= \begin{pmatrix} 0 & -I_n\\ I_n & 0\end{pmatrix}.\end{aligned}$$

Thus if $M \in Sp(n,\mathbb{R})$, $\det M = 1$.

A subgroup Γ is said to be *discrete* if $\Gamma \cap K$ is finite for any compact subset K of $Sp(n,\mathbb{R})$.

Theorem 13. *A discrete subgroup Γ of $Sp(n,\mathbb{R})$ acts properly discontinuously on $\mathbb{H}_n$, that is, for any two compact subsets C_1, C_2 of $\mathbb{H}_n$, the set*

$$\{\,\gamma \in \Gamma \mid \gamma\cdot C_1 \cap C_2 \neq \emptyset\,\}$$

is finite.

Proof. We can show that the mapping

$$p : Sp(n,\mathbb{R}) \longrightarrow \mathbb{H}_n, \qquad M \longrightarrow M\cdot(iI_n), \quad M \in Sp(n,\mathbb{R})$$

is *proper*, i.e., for any compact subset $X \subset \mathbb{H}_n$, $p^{-1}(X)$ is compact in $Sp(n,\mathbb{R})$ (cf. [7], pp. 28-29). Suppose X_1 and X_2 are two compact subsets of $\mathbb{H}_n$. Then $Z_1 = p^{-1}(X_1)$ and $Z_2 = p^{-1}(X_2)$ are compact in $Sp(n,\mathbb{R})$. Since the image of $Z_2 \times Z_1$ under the continuous mapping $(M_2, M_1) \mapsto M_2M_1^{-1}$ is compact, the set

$$Z_2Z_1^{-1} = \{\, M_2M_1^{-1} \mid M_1 \in Z_1,\ M_2 \in Z_2 \,\}$$

is compact. It remains to show that $\{\gamma \in \Gamma \mid \gamma\cdot X_1 \cap X_2 \neq \emptyset\}$ is finite. If $\gamma \in \Gamma$ such that $\gamma\cdot X_1 \cap X_2 \neq \emptyset$, then

$$\gamma\cdot\Omega_1 = M_2\cdot(iI_n) \in X_2 \quad \text{for some } \Omega_1 \in X_1 \ \text{ and } M_2 \in Z_2.$$

Since $(\gamma^{-1}M_2)\cdot(iI_n) = \Omega_1 \in X_1$, we have $\gamma^{-1}M_2 \in Z_1$, that is, $M_2^{-1}\gamma \in Z_1^{-1}$. Therefore $\gamma \in M_2Z_1^{-1} \subset Z_2Z_1^{-1}$. Since $\Gamma \cap Z_2Z_1^{-1}$ is finite, the set $\{\gamma \in \Gamma \mid \gamma\cdot X_1 \cap X_2 \neq \emptyset\}$ is finite. □

By Theorem 13, the Siegel modular group $\Gamma_n = Sp(n,\mathbb{Z})$ acts properly discontinuously on $\mathbb{H}_n$. Therefore the stabilizer $(\Gamma_n)_\Omega$ of $\Omega \in \mathbb{H}_n$ given by

$$(\Gamma_n)_\Omega = \{\gamma \in \Gamma_n \mid \gamma\cdot\Omega = \Omega\}$$

is a finite subgroup of Γ_n.

Let q be a positive integer. The set

$$\Gamma_n(q) = \{\, M \in \Gamma_n \mid M \equiv I_{2n} \bmod q \,\}$$

is a normal subgroup of Γ_n because it is the kernel of the homomorphism $\Gamma_n \longrightarrow Sp(n,\mathbb{Z}/q\mathbb{Z})$ defined by $\gamma \longrightarrow \gamma \bmod q$. It is called the *principal congruence subgroup* of *level* q. We have $\Gamma_n(1) = \Gamma_n$. A subgroup Γ of $Sp(n,\mathbb{R})$ which contains $\Gamma_n(\ell)$ for some positive integer ℓ as a subgroup of finite index is called a *modular group*. A subgroup Γ of Γ_n which contains $\Gamma_n(\ell)$ for some positive integer ℓ as a subgroup of finite index is called a *congruence subgroup* of Γ_n. The subset $\Gamma_{\vartheta,n}$ of Γ_n consisting of elements $\gamma = \begin{pmatrix} A & B \\ C & D \end{pmatrix} \in \Gamma_n$ such that the diagonal elements of A^tB and C^tD are even integers is a subgroup of Γ_n called the *theta group*. For a positive integer q, we let

$$\Gamma_{n,0}(q) = \left\{ \begin{pmatrix} A & B \\ C & D \end{pmatrix} \in \Gamma_n \,\middle|\, C \equiv 0 \bmod q \right\}.$$

Then $\Gamma_{n,0}(q)$ is a congruence subgroup of Γ_n containing the principal congruence subgroup $\Gamma_n(q)$ of level q.

Let Ω_1 and Ω_2 be two points of $\mathbb{H}_n$ and $M = \begin{pmatrix} A & B \\ C & D \end{pmatrix} \in Sp(n,\mathbb{R})$. We write $\Omega_i^* = M\cdot\Omega_i$ $(i = 1,2)$. Then by the symplectic conditions (2.1.1) and

(2.1.2), we have

$$\Omega_2^* - \Omega_1^* = {}^t(C\Omega_2 + D)^{-1}(\Omega_2 - \Omega_1)(C\Omega_1 + D)^{-1} \tag{2.1.7}$$

and

$$\Omega_2^* - \overline{\Omega_1^*} = {}^t(C\Omega_2 + D)^{-1}\big(\Omega_2 - \overline{\Omega_1}\big)(C\overline{\Omega_1} + D)^{-1} \tag{2.1.8}$$

Let $\Omega = X + i\,Y \in \mathbb{H}_n$ with $X, Y \in \mathbb{R}^{(n,n)}$. If $\Omega^* = M\cdot\Omega$, then we write $\Omega^* = X^* + i\,Y^*$ with $X^*, Y^* \in \mathbb{R}^{(n,n)}$. Then by (2.1.8),

$$\Omega^* - \overline{\Omega^*} = {}^t(C\Omega + D)^{-1}(\Omega - \overline{\Omega}\,)(C\overline{\Omega} + D)^{-1} \tag{2.1.9}$$

and hence

$$Y^* = {}^t(C\Omega + D)^{-1}Y(C\overline{\Omega} + D)^{-1}. \tag{2.1.10}$$

Therefore we obtain

$$\det Y^* = \det Y \cdot |\det(C\Omega + D)|^{-2}. \tag{2.1.11}$$

And

$$\begin{aligned} d\Omega^* &= d(M\cdot\Omega) = d\big\{(A\Omega + B)(C\Omega + D)^{-1}\big\} \\ &= A\,d\Omega\,(C\Omega + D)^{-1} - (A\Omega + B)(C\Omega + D)^{-1}C\,d\Omega\,(C\Omega + D)^{-1} \\ &= {}^t(C\Omega + D)^{-1}\big\{\Omega({}^tCA - {}^tAC) + ({}^tDA - {}^tBC)\big\}d\Omega\,(C\Omega + D)^{-1} \\ &= {}^t(C\Omega + D)^{-1}d\Omega\,(C\Omega + D)^{-1}. \end{aligned}$$

Thus we have

$$d\Omega^* = {}^t(C\Omega + D)^{-1}\,d\Omega\,(C\Omega + D)^{-1}. \tag{2.1.12}$$

By Formulas (2.1.10) and (2.1.12),

$$ds^2 = \sigma(Y^{-1}d\Omega\,Y^{-1}d\overline{\Omega})$$

is invariant under the action (2.1.6) of $Sp(n,\mathbb{R})$. For $\Omega = iI_n$,

$$ds^2 = \sum_{i=1} \big(dx_{ii}^2 + dy_{ii}^2\big) + 2 \sum_{1\le i\lneq j\le n} \big(dx_{ij}^2 + dy_{ij}^2\big).$$

Since $Sp(n,\mathbb{R})$ acts on $\mathbb{H}_n$ transitively, ds^2 is an $Sp(n,\mathbb{R})$-invariant Riemannian metric on $\mathbb{H}_n$.

The tangent space $T_\Omega(\mathbb{H}_n)$ of $\mathbb{H}_n$ at Ω is identified with the vector space $\mathrm{Sym}(n,\mathbb{C})$ consisting of all $n\times n$ symmetric complex matrices (cf. (2.2.20) in Section 2.2). By (2.1.12), the differential

$$dM_\Omega : T_\Omega(\mathbb{H}_n) \longrightarrow T_{M\cdot\Omega}(\mathbb{H}_n)$$

of the symplectic transformation M at Ω is given by

$$dM_\Omega(W) = {}^t(C\Omega + D)^{-1}\, W\, (C\Omega + D)^{-1}, \quad W \in \mathrm{Sym}(n, \mathbb{C}). \qquad (2.1.13)$$

We can see that the Jacobian of the symplectic transformation $M \in Sp(n, \mathbb{C})$ is given by

$$\frac{\partial(\Omega^*)}{\partial(\Omega)} = \det(C\Omega + D)^{-(n+1)},$$

where $\Omega^* = M \cdot \Omega$ with $M = \begin{pmatrix} A & B \\ C & D \end{pmatrix} \in Sp(n, \mathbb{R})$.

Finally we describe the universal covering group of $Sp(n, \mathbb{R})$ using the so-called Maslov index. Let (V, B) be a symplectic (real) vector space of dimension $2n$ with a non-degenerate alternating form B on V. A subspace of (V, B) such that $B(x, y) = 0$ for all $x, y \in L$ is said to be totally isotropic. For a subspace L of (V, B), we will denote by $L^\perp$ the orthogonal complement of L in V relative to B, i.e.,

$$L^\perp = \{ x \in V \mid B(x, y) = 0 \quad \text{for all } y \in V \}.$$

If L is a subspace of (V, B) such that $L = L^\perp$, then L is called a Lagrangian subspace of (V, B). If L is a totally isotropic subspace of V such that $B(x, L) = 0$ implies $x \in L$, then L is said to be maximally totally isotropic. We note that if L is a Lagrangian subspace of (V, B), then $\dim L = n$ because $\dim L + \dim L^\perp = 2n$.

Let $Sp(B)$ be the symplectic group defined by

$$Sp(B) = \{ g \in GL(V) \mid B(gx, gy) = B(x, y) \quad \text{for all } x, y \in V \}.$$

Definition 5. Let L_1, L_2, L_3 be three Lagrangian subspaces of V. The integer $\tau(L_1, L_2, L_3)$ is defined to be the signature of the quadratic form $Q(x_1 + x_2 + x_3)$ on the $3n$-dimensional vector space $L_1 \oplus L_2 \oplus L_3$ defined by

$$Q(x_1 + x_2 + x_3) := B(x_1, x_2) + B(x_2, x_3) + B(x_3, x_1),$$

where $x_1 \in L_1$, $x_2 \in L_2$ and $x_3 \in L_3$. The integer $\tau(L_1, L_2, L_3)$ is called the Maslov index of (V, B).

Lemma 30. *Let L_1, L_2, L_3 be three Lagrangian subspaces of (V, B). Then we have the following properties:*

(1) The Maslov index is $Sp(n, \mathbb{R})$-invariant, i.e., for any $g \in Sp(n, \mathbb{R})$, we have $\tau(gL_1, gL_2, gL_3) = \tau(L_1, L_2, L_3)$.

(2) $\tau(L_1, L_2, L_3) = -\tau(L_2, L_1, L_3) = -\tau(L_1, L_3, L_2)$.

Proof. It follows immediately from the definition. □

For a sequence $(L_1, L_2, \cdots, L_k)$ with $k \geq 4$ of Lagrangian subspaces of (V, B), we define the **generalized Maslov index** $\tau(L_1, L_2, \cdots, L_k)$ by

$$\tau(L_1, L_2, \cdots, L_k) = \tau(L_1, L_2, L_3) + \tau(L_1, L_3, L_4) + \cdots + \tau(L_1, L_{k-1}, L_k).$$

Proposition 8. *(1) The Maslov index $\tau(L_1, L_2, \cdots, L_k)$ is invariant under the action of the symplectic group $Sp(B)$, and its value is unchanged under circular permutation.*
(2) For any Lagrangian subspace $L_1, L_2, L_3, L_1', L_2', L_3'$, we have

$$\begin{aligned}\tau(L_1', L_2', L_3') = &\ \tau(L_1, L_2, L_3) + \tau(L_1', L_2', L_2, L_1) + \tau(L_2', L_3', L_3, L_2) \\ &+ \tau(L_3', L_1', L_1, L_3).\end{aligned}$$

Proof. See [21, pp. 45-46]. □

Let Λ be the space of all Lagrangian subspaces of (V, B). Then Λ may be regarded as a closed submanifold of the Grassmannian manifold of all n-dimensional subspaces in $\mathbb{R}^{2n}$. We define

$$\widetilde{\Lambda} := \Lambda \times \mathbb{Z} = \{\, (L, u) \,|\, L \in \Lambda,\ u \in \mathbb{Z} \,\}.$$

We fix a Lagrangian subspace L_0 of (V, B). Let $(L_1, u_1) \in \widetilde{\Lambda}$ and let $\mathscr{U}$ be a neighborhood of L_1. Let L_2 be a Lagrangian subspace of V transverse to L_1. We define

$$U(L_1, u_1; \mathscr{U}, L_2) := \{\, (L, u) \,|\, L \in \mathscr{U},\ u = u_1 + \tau(L, L_0, L_1, L_2) \,\}.$$

It is proved in Proposition 1.9.5 in [21] that the set of all such $U(L_1, u_1; \mathscr{U}, L_2)$'s form a neighborhood for a topology on $\widetilde{\Lambda}$. Let $\pi : \widetilde{\Lambda} \longrightarrow \Lambda$ be the projection defined by $\pi(L, u) = L$. Clearly π is a continuous map and hence $\widetilde{\Lambda}$ is a covering of Λ.

Let L_* be a fixed element of Λ. We define the group

$$\widetilde{Sp(B)}_* := Sp(B) \times \mathbb{Z} \tag{2.1.14}$$

equipped with the multiplication law

$$(g_1, n_1) \cdot (g_2, n_2) = \big(g_1 g_2, n_1 + n_2 + \tau(L_*, g_1 L_*, g_1 g_2 L_*)\big),$$

where $g_1, g_2 \in Sp(B)$ and $n_1, n_2 \in \mathbb{Z}$. Then it is easy to see that $\widetilde{Sp(B)}_*$ acts on $\widetilde{\Lambda}$ by

$$(g, n) \cdot (L, u) = \big(gL, n + u + \tau(L_*, gL_*, gL)\big),$$

where $g \in Sp(B),\ n, u \in \mathbb{Z}$ and $L \in \Lambda$.

Let L_2 be a Lagrangian subspace of (V, B) transverse to L_* and $\mathscr{E}$ be a neighborhood of e in $Sp(B)$, where e is the identity element of $Sp(B)$. We define

$$\mathscr{W}(\mathscr{E}, L_2) := \{ (g, -\tau(gL_*, L_2, L_*)) \mid g \in \mathscr{E} \}.$$

Then the set of all such $\mathscr{W}(\mathscr{E}, L_2)$'s form a fundamental system of neighborhoods of $(e, 0)$ on $\widetilde{Sp(B)}_*$. Therefore $\widetilde{Sp(B)}_*$ has the structure of a topological group. It is easily seen that $\widetilde{Sp(B)}_*$ acts on $\widetilde{\Lambda}$ continuously.

Definition 6. An *oriented* vector space of dimension n is a pair (W, ε), where W is a real vector space of dimension n and ε is an orientation of W, i.e., a connected component of $\Lambda^n L - \{0\}$. If (W_1, ε_1) and (W_2, ε_2) are two oriented vector spaces of dimension n and A is a linear invertible map from W_1 to W_2, we define the sign of the determinant of A denoted by $\delta(A) = \pm 1$, by the condition

$$(\Lambda^n A)\, \varepsilon_1 = c\, \delta(A)\, \varepsilon_2 \quad \text{with } c > 0.$$

L and M be two Lagrangian subspaces of a symplectic vector space (V, B). We define $g_{M,L} : L \longrightarrow M^*$ by $\langle g_{M,L}(x), y \rangle = B(x, y)$ for all $x \in L$ and $y \in M$. Here M^* denotes the dual vector space of M. Let (L_1, ε_1) and (L_2, ε_2) be two oriented Lagrangian subspaces of (V, B) which are transverse. Then $g_{L_2, L_1} : (L_1, \varepsilon_1) \longrightarrow (L_2, \varepsilon_2)$. We define

$$\xi\big((L_1, \varepsilon_1), (L_2, \varepsilon_2)\big) := \delta(g_{L_2, L_1}).$$

This depends only on the relative orientation of (L_1, ε_1) and (L_2, ε_2). More generally if L_1 and L_2 are not transverse, we define (L_1, ε_1) and (L_2, ε_2) as follows: Let ε be an orientation of $H = L_1 \cap L_2$. Then ε defines an orientation $\widetilde{\varepsilon}_i$ $(i = 1, 2)$ on L_i/H by $\widetilde{\varepsilon}_i \wedge \varepsilon = \varepsilon_i$. Since L_1/H and L_2/H are two transverse subspaces of $(L_1 + L_2)/H = H^{\perp}/H$, we can define

$$\xi\big((L_1, \varepsilon_1), (L_2, \varepsilon_2)\big) := \xi\big((L_1/H, \widetilde{\varepsilon}_1), (L_2/H, \widetilde{\varepsilon}_2)\big).$$

We observe that this is independent of the choice of the orientation ε of H because $\widetilde{\varepsilon}_1$ and $\widetilde{\varepsilon}_2$ change simultaneously if we change ε to $-\varepsilon$.

If $L_1 = L_2$, we define

$$\xi\big((L_1, \varepsilon_1), (L_2, \varepsilon_2)\big) := \begin{cases} 1 & \text{if } \varepsilon_1 = \varepsilon_2, \\ -1 & \text{if } \varepsilon_1 \neq \varepsilon_2. \end{cases}$$

Definition 7. Let (L_1, ε_1) and (L_2, ε_2) be two oriented Lagrangian subspaces of a symplectic vector space (V, B). We define

$$s\big((L_1, \varepsilon_1), (L_2, \varepsilon_2)\big) := \big(\sqrt{-1}\big)^{n - \dim(L_1 \cap L_2)}\, \xi\big((L_1, \varepsilon_1), (L_2, \varepsilon_2)\big).$$

Definition 8. Let L be a Lagrangian subspace of a symplectic vector space (V, B). We choose an orientation L^+ on L. We define the map $s_L : Sp(B) \longrightarrow \mathbb{C}$ by

$$s_L(g) := s(L^+, gL^+), \quad g \in Sp(B).$$

This is well-defined because $s_L(g)$ is independent of the choice of the orientation on L.

We define the map $\tilde{s}_* : \widetilde{Sp(B)}_* \longrightarrow \mathbb{C}$ by

$$\tilde{s}_*(g,n) := e^{\frac{\pi n i}{2}} s_{L_*}(g), \quad g \in Sp(B),\ n \in \mathbb{Z}.$$

Lemma 31. *$\tilde{s}_*(g,n)$ is a character of $\widetilde{Sp(B)}_*$ with values in $\mathbb{Z}/4\mathbb{Z}$.*

Proof. The proof can be found in [21, p. 72]. □

We see that the kernel of $\tilde{s}_*$ is the universal covering group of $Sp(B)$ and the fundamental group $\pi_1(Sp(B))$ of $Sp(B)$ is isomorphic to $\mathbb{Z}$. Therefore $\widetilde{Sp(B)}_*$ is the union of four connected components such that each of them is simply connected.

We now consider the group

$$Sp(B, L_*) := Sp(B) \times \mathbb{C}_1^*$$

equipped with the multiplication law

$$(g_1, t_1) \cdot (g_2, t_2) = \big(g_1 g_2,\ t_1 t_2\, c_*(g_1, g_2) \big),$$

where $g_1, g_2 \in Sp(B),\ t_1, t_2 \in \mathbb{C}_1^*$ and

$$c_*(g_1, g_2) := e^{-\frac{\pi i}{4}\, \tau(L_*, g_1 L_*, g_1 g_2 L_*)}.$$

It is easily checked that the $\varphi : \widetilde{Sp(B)}_* \longrightarrow Sp(B, L_*)$ defined by

$$\varphi(g,n) := \big(g, e^{\frac{\pi n i}{4}}\big), \quad g \in Sp(B),\ n \in \mathbb{Z}$$

is a group homomorphism. We define

$$Mp(B)_* := \big\{ (g,t) \in Sp(B, L_*) \,\big|\, t^2 = s_{L_*}(g)^{-1} \big). \tag{2.1.15}$$

2.2 Some Geometry on Siegel Space

For $\Omega = (\omega_{ij}) \in \mathbb{H}_n$, we write $\Omega = X + iY$ with $X = (x_{ij}),\ Y = (y_{ij})$ real and $d\Omega = (d\omega_{ij})$. We put

$$\frac{\partial}{\partial\Omega} = \left(\frac{1+\delta_{ij}}{2}\,\frac{\partial}{\partial\omega_{ij}}\right) \quad \text{and} \quad \frac{\partial}{\partial\overline{\Omega}} = \left(\frac{1+\delta_{ij}}{2}\,\frac{\partial}{\partial\overline{\omega}_{ij}}\right).$$

C. L. Siegel [35] introduced the symplectic metric ds^2 on $\mathbb{H}_n$ invariant under the action (2.1.5) of $Sp(n,\mathbb{R})$ given by

$$ds^2 = \sigma(Y^{-1}d\Omega\, Y^{-1}d\overline{\Omega}) \tag{2.2.1}$$

and H. Maass [22] proved that its Laplacian is given by

$$\Delta = 4\,\sigma\left(Y\,{}^t\!\left(Y\frac{\partial}{\partial\overline{\Omega}}\right)\frac{\partial}{\partial\Omega}\right). \tag{2.2.2}$$

And

$$dv_n(\Omega) = (\det Y)^{-(n+1)} \prod_{1\le i\le j\le n} dx_{ij} \prod_{1\le i\le j\le n} dy_{ij} \tag{2.2.3}$$

is a $Sp(n,\mathbb{R})$-invariant volume element on $\mathbb{H}_n$ (cf. [37], p. 130).

Theorem 14. (**Siegel [35]**). *(1) There exists exactly one geodesic joining two arbitrary points Ω_0, Ω_1 in $\mathbb{H}_n$. Let $R(\Omega_0,\Omega_1)$ be the cross-ratio defined by*

$$R(\Omega_0,\Omega_1) = (\Omega_0-\Omega_1)(\Omega_0-\overline{\Omega}_1)^{-1}(\overline{\Omega}_0-\overline{\Omega}_1)(\overline{\Omega}_0-\Omega_1)^{-1}. \tag{2.2.4}$$

For brevity, we put $R_* = R(\Omega_0,\Omega_1)$. Then the symplectic length $\rho(\Omega_0,\Omega_1)$ of the geodesic joining Ω_0 and Ω_1 is given by

$$\rho(\Omega_0,\Omega_1)^2 = \sigma\left(\left(\log\frac{1+R_*^{\frac12}}{1-R_*^{\frac12}}\right)^2\right), \tag{2.2.5}$$

where

$$\left(\log\frac{1+R_*^{\frac12}}{1-R_*^{\frac12}}\right)^2 = 4\,R_*\left(\sum_{k=0}^{\infty}\frac{R_*^k}{2k+1}\right)^2.$$

(2) For $M \in Sp(n,\mathbb{R})$, we set

$$\widetilde{\Omega}_0 = M\cdot\Omega_0 \quad \textit{and} \quad \widetilde{\Omega}_1 = M\cdot\Omega_1.$$

Then $R(\Omega_1,\Omega_0)$ and $R(\widetilde{\Omega}_1,\widetilde{\Omega}_0)$ have the same eigenvalues.
(3) All geodesics are symplectic images of the special geodesics

$$\alpha(t) = i\,\mathrm{diag}(a_1^t, a_2^t, \cdots, a_n^t), \tag{2.2.6}$$

where $a_1, a_2, \cdots, a_n$ *are arbitrary positive real numbers satisfying the condition*

$$\sum_{k=1}^{n} (\log a_k)^2 = 1.$$

The proof of the above theorem can be found in [35], pp. 289-293.

Let

$$\mathbb{D}_n = \left\{ W \in \mathbb{C}^{(n,n)} \mid W = {}^tW,\ I_n - W\overline{W} > 0 \right\}$$

be the generalized unit disk of degree n. The Cayley transform $\Psi : \mathbb{D}_n \longrightarrow \mathbb{H}_n$ defined by

$$\Psi(W) = i\,(I_n + W)(I_n - W)^{-1}, \quad W \in \mathbb{D}_n \tag{2.2.7}$$

is a biholomorphic mapping of $\mathbb{D}_n$ onto $\mathbb{H}_n$ which gives the bounded realization of $\mathbb{H}_n$ by $\mathbb{D}_n$ (cf. [35]). A. Korányi and J. Wolf [20] gave a realization of a bounded symmetric domain as a Siegel domain of the third kind investigating a generalized Cayley transform of a bounded symmetric domain that generalizes the Cayley transform Ψ of $\mathbb{D}_n$.

Let

$$T = \frac{1}{\sqrt{2}} \begin{pmatrix} I_n & I_n \\ iI_n & -iI_n \end{pmatrix} \tag{2.2.8}$$

be the $2n \times 2n$ matrix represented by Ψ. Then

$$T^{-1}Sp(n,\mathbb{R})\,T = \left\{ \begin{pmatrix} P & Q \\ \overline{Q} & \overline{P} \end{pmatrix} \;\middle|\; {}^tP\overline{P} - {}^t\overline{Q}Q = I_n,\ {}^tP\overline{Q} = {}^t\overline{Q}P \right\}. \tag{2.2.9}$$

Indeed, if $M = \begin{pmatrix} A & B \\ C & D \end{pmatrix} \in Sp(n,\mathbb{R})$, then

$$T^{-1}MT = \begin{pmatrix} P & Q \\ \overline{Q} & \overline{P} \end{pmatrix}, \tag{2.2.10}$$

where

$$P = \frac{1}{2}\left\{ (A+D) + i\,(B-C) \right\} \tag{2.2.11}$$

and

$$Q = \frac{1}{2}\left\{ (A-D) - i\,(B+C) \right\}. \tag{2.2.12}$$

For brevity, we set

$$G_* = T^{-1}Sp(n,\mathbb{R})T.$$

Then G_* is a subgroup of $SU(n,n)$, where

$$SU(n,n)=\left\{\,h\in\mathbb{C}^{(n,n)}\mid {}^th I_{n,n}\overline{h}=I_{n,n}\,\right\},\quad I_{n,n}=\begin{pmatrix} I_n & 0\\ 0 & -I_n\end{pmatrix}.$$

In the case $n=1$, we observe that

$$T^{-1}Sp(1,\mathbb{R})T=T^{-1}SL_2(\mathbb{R})T=SU(1,1).$$

If $n>1$, then G_* is a *proper* subgroup of $SU(n,n)$. In fact, since ${}^tTJ_nT=-i\,J_n$, we get

$$G_*=\left\{\,h\in SU(n,n)\,|\,{}^thJ_nh=J_n\,\right\}=SU(n,n)\cap Sp(n,\mathbb{C}),\qquad (2.2.13)$$

where

$$Sp(n,\mathbb{C})=\left\{\alpha\in\mathbb{C}^{(2n,2n)}\mid {}^t\alpha\,J_n\,\alpha=J_n\,\right\}.$$

Let

$$P^+=\left\{\begin{pmatrix} I_n & Z\\ 0 & I_n\end{pmatrix}\,\Big|\, Z={}^tZ\in\mathbb{C}^{(n,n)}\right\}$$

be the P^+-part of the complexification of $G_*\subset SU(n,n)$. We note that the Harish-Chandra decomposition of an element $\begin{pmatrix} P & Q\\ \overline{Q} & \overline{P}\end{pmatrix}$ in G_* is

$$\begin{pmatrix} P & Q\\ \overline{Q} & \overline{P}\end{pmatrix}=\begin{pmatrix} I_n & Q\overline{P}^{-1}\\ 0 & I_n\end{pmatrix}\begin{pmatrix} P-Q\overline{P}^{-1}\overline{Q} & 0\\ 0 & \overline{P}\end{pmatrix}\begin{pmatrix} I_n & 0\\ \overline{P}^{-1}\overline{Q} & I_n\end{pmatrix}.$$

For more detail, we refer to [19, p. 155]. Thus the P^+-component of the following element

$$\begin{pmatrix} P & Q\\ \overline{Q} & \overline{P}\end{pmatrix}\cdot\begin{pmatrix} I_n & W\\ 0 & I_n\end{pmatrix},\quad W\in\mathbb{D}_n$$

of the complexification of G_*^J is given by

$$\begin{pmatrix} I_n & (PW+Q)(\overline{Q}W+\overline{P})^{-1}\\ 0 & I_n\end{pmatrix}.\qquad (2.2.14)$$

We note that $Q\overline{P}^{-1}\in\mathbb{D}_n$. We get the Harish-Chandra embedding of $\mathbb{D}_n$ into P^+ (cf. [19, p. 155] or [33, pp. 58-59]). Therefore we see that G_* acts on $\mathbb{D}_n$ transitively by

$$\begin{pmatrix} P & Q\\ \overline{Q} & \overline{P}\end{pmatrix}\cdot W=(PW+Q)(\overline{Q}W+\overline{P})^{-1},\quad \begin{pmatrix} P & Q\\ \overline{Q} & \overline{P}\end{pmatrix}\in G_*,\ W\in\mathbb{D}_n.\quad (2.2.15)$$

The isotropy subgroup K_* of G_* at the origin o is given by

$$K_*=\left\{\begin{pmatrix} P & 0\\ 0 & \overline{P}\end{pmatrix}\,\Big|\, P\in U(n)\right\}.$$

Thus G_*/K_* is biholomorphic to $\mathbb{D}_n$. It is known that the action (2.1.6) is compatible with the action (2.2.15) via the Cayley transform Ψ (cf. (2.2.7)). In other words, if $M \in Sp(n,\mathbb{R})$ and $W \in \mathbb{D}_n$, then

$$M \cdot \Psi(W) = \Psi(M_* \cdot W), \tag{2.2.16}$$

where $M_* = T^{-1}MT \in G_*$.

For $W = (w_{ij}) \in \mathbb{D}_n$, we write $dW = (dw_{ij})$ and $d\overline{W} = (d\overline{w}_{ij})$. We put

$$\frac{\partial}{\partial W} = \left(\frac{1+\delta_{ij}}{2} \frac{\partial}{\partial w_{ij}} \right) \quad \text{and} \quad \frac{\partial}{\partial \overline{W}} = \left(\frac{1+\delta_{ij}}{2} \frac{\partial}{\partial \overline{w}_{ij}} \right).$$

Using the Cayley transform $\Psi : \mathbb{D}_n \longrightarrow \mathbb{H}_n$, Siegel showed (cf. [35]) that

$$ds_*^2 = 4\sigma\Big((I_n - W\overline{W})^{-1} dW\, (I_n - \overline{W}W)^{-1} d\overline{W}\Big) \tag{2.2.17}$$

is a G_*-invariant Riemannian metric on $\mathbb{D}_n$ and Maass [22] showed that its Laplacian is given by

$$\Delta_* = \sigma\left((I_n - W\overline{W})\, {}^t\!\left((I_n - W\overline{W}) \frac{\partial}{\partial \overline{W}} \right) \frac{\partial}{\partial W} \right). \tag{2.2.18}$$

Now we discuss the differential operators on $\mathbb{H}_n$ invariant under the action (2.1.6). The isotropy subgroup K at iI_n for the action (2.1.6) is a maximal compact subgroup given by

$$K = \left\{ \begin{pmatrix} A & -B \\ B & A \end{pmatrix} \,\Big|\, A\,{}^tA + B\,{}^tB = I_n,\ A\,{}^tB = B\,{}^tA,\ A, B \in \mathbb{R}^{(n,n)} \right\}.$$

Let $\mathfrak{k}$ be the Lie algebra of K. Then the Lie algebra $\mathfrak{sp}(n,\mathbb{R})$ of $Sp(n,\mathbb{R})$ has a Cartan decomposition $\mathfrak{sp}(n,\mathbb{R}) = \mathfrak{k} \oplus \mathfrak{p}$, where

$$\mathfrak{p} = \left\{ \begin{pmatrix} X & Y \\ Y & -X \end{pmatrix} \,\Big|\, X = {}^tX,\ Y = {}^tY,\ X, Y \in \mathbb{R}^{(n,n)} \right\}.$$

The subspace $\mathfrak{p}$ of $\mathfrak{sp}(n,\mathbb{R})$ may be regarded as the tangent space of $\mathbb{H}_n$ at iI_n. The adjoint representation of $Sp(n,\mathbb{R})$ on $\mathfrak{sp}(n,\mathbb{R})$ induces the action of K on $\mathfrak{p}$ given by

$$k \cdot Z = kZ\,{}^tk, \quad k \in K,\ Z \in \mathfrak{p}. \tag{2.2.19}$$

Let $\mathbb{T}_n$ be the vector space of $n \times n$ symmetric complex matrices. We let $\psi : \mathfrak{p} \longrightarrow \mathbb{T}_n$ be the map defined by

$$\psi\left(\begin{pmatrix} X & Y \\ Y & -X \end{pmatrix} \right) = X + i\,Y, \quad \begin{pmatrix} X & Y \\ Y & -X \end{pmatrix} \in \mathfrak{p}. \tag{2.2.20}$$

We let $\delta : K \longrightarrow U(n)$ be the isomorphism defined by

$$\delta\left(\begin{pmatrix} A & -B \\ B & A \end{pmatrix}\right) = A + i\,B, \quad \begin{pmatrix} A & -B \\ B & A \end{pmatrix} \in K, \tag{2.2.21}$$

where $U(n)$ denotes the unitary group of degree n. We identify $\mathfrak{p}$ (resp. K) with $\mathbb{T}_n$ (resp. $U(n)$) through the map Ψ (resp. δ). We consider the action of $U(n)$ on $\mathbb{T}_n$ defined by

$$h \cdot Z = hZ\,{}^t h, \quad h \in U(n),\ Z \in \mathbb{T}_n. \tag{2.2.22}$$

Then the adjoint action (2.2.19) of K on $\mathfrak{p}$ is compatible with the action (2.2.22) of $U(n)$ on $\mathbb{T}_n$ through the map ψ. Precisely for any $k \in K$ and $\omega \in \mathfrak{p}$, we get

$$\psi(k\,\omega\,{}^t k) = \delta(k)\,\psi(\omega)\,{}^t\delta(k). \tag{2.2.23}$$

The action (2.2.22) induces the action of $U(n)$ on the polynomial algebra $\mathrm{Pol}(\mathbb{T}_n)$ and the symmetric algebra $S(\mathbb{T}_n)$ respectively. We denote by $\mathrm{Pol}(\mathbb{T}_n)^{U(n)}$ $\left(\text{resp. } S(\mathbb{T}_n)^{U(n)}\right)$ the subalgebra of $\mathrm{Pol}(\mathbb{T}_n)$ $\left(\text{resp. } S(\mathbb{T}_n)\right)$ consisting of $U(n)$-invariants. The following inner product $(\ ,\ )$ on $\mathbb{T}_n$ defined by

$$(Z, W) = \mathrm{tr}\big(Z\,\overline{W}\,\big), \quad Z, W \in \mathbb{T}_n$$

gives an isomorphism as vector spaces

$$\mathbb{T}_n \cong \mathbb{T}_n^*, \quad Z \mapsto f_Z, \quad Z \in \mathbb{T}_n, \tag{2.2.24}$$

where $\mathbb{T}_n^*$ denotes the dual space of $\mathbb{T}_n$ and f_Z is the linear functional on $\mathbb{T}_n$ defined by

$$f_Z(W) = (W, Z), \quad W \in \mathbb{T}_n.$$

It is known that there is a canonical linear bijection of $S(\mathbb{T}_n)^{U(n)}$ onto the algebra $\mathbb{D}(\mathbb{H}_n)$ of differential operators on $\mathbb{H}_n$ invariant under the action (2.1.6) of $Sp(n, \mathbb{R})$. Identifying $\mathbb{T}_n$ with $\mathbb{T}_n^*$ by the above isomorphism (2.2.24), we get a canonical linear bijection

$$\Phi : \mathrm{Pol}(\mathbb{T}_n)^{U(n)} \longrightarrow \mathbb{D}(\mathbb{H}_n) \tag{2.2.25}$$

of $\mathrm{Pol}(\mathbb{T}_n)^{U(n)}$ onto $\mathbb{D}(\mathbb{H}_n)$. The map Φ is described explicitly as follows. Similarly the action (2.2.19) induces the action of K on the polynomial algebra $\mathrm{Pol}(\mathfrak{p})$ and $S(\mathfrak{p})$ respectively. Through the map ψ, the subalgebra $\mathrm{Pol}(\mathfrak{p})^K$ of $\mathrm{Pol}(\mathfrak{p})$ consisting of K-invariants is isomorphic to $\mathrm{Pol}(\mathbb{T}_n)^{U(n)}$. We put $N = n(n+1)$. Let $\{\xi_\alpha \mid 1 \le \alpha \le N\}$ be a basis of $\mathfrak{p}$. If $P \in \mathrm{Pol}(\mathfrak{p})^K$, then

$$\big(\Phi(P)f\big)(gK) = \left[P\left(\frac{\partial}{\partial t_\alpha}\right) f\left(g \exp\left(\sum_{\alpha=1}^N t_\alpha \xi_\alpha\right) K\right)\right]_{(t_\alpha)=0}, \tag{2.2.26}$$

where $f \in C^\infty(\mathbb{H}_n)$ and $g \in Sp(n,\mathbb{R})$. We refer to [11, 12] for more detail. In general, it is hard to express $\Phi(P)$ explicitly for a polynomial $P \in \mathrm{Pol}(\mathfrak{p})^K$.

According to the work of Harish-Chandra [9, 10], the algebra $\mathbb{D}(\mathbb{H}_n)$ is generated by n algebraically independent generators and is isomorphic to the commutative ring $\mathbb{C}[x_1,\cdots,x_n]$ with n indeterminates. We note that n is the real rank of $Sp(n,\mathbb{R})$. Let $\mathfrak{sp}(n,\mathbb{C})$ be the complexification of $\mathfrak{sp}(n,\mathbb{R})$. It is known that $\mathbb{D}(\mathbb{H}_n)$ is isomorphic to the center of the universal enveloping algebra of $\mathfrak{sp}(n,\mathbb{C})$ (cf. [34]).

Using a classical invariant theory (cf. [13, 40]), we can show that $\mathrm{Pol}(\mathbb{T}_n)^{U(n)}$ is generated by the following algebraically independent polynomials

$$q_j(Z) = \sigma\Big(\big(Z\overline{Z}\big)^j\Big), \quad j = 1,2,\cdots,n. \tag{2.2.27}$$

For each j with $1 \le j \le n$, the image $\Phi(q_j)$ of q_j is an invariant differential operator on $\mathbb{H}_n$ of degree $2j$. The algebra $\mathbb{D}(\mathbb{H}_n)$ is generated by n algebraically independent generators $\Phi(q_1),\Phi(q_2),\cdots,\Phi(q_n)$. In particular,

$$\Phi(q_1) = c_1\,\sigma\left(Y\,{}^t\!\left(Y\frac{\partial}{\partial\overline{\Omega}}\right)\frac{\partial}{\partial\Omega}\right) \quad \text{for some constant } c_1. \tag{2.2.28}$$

We observe that if we take $Z = X + i\,Y$ with real X, Y, then $q_1(Z) = q_1(X,Y) = \sigma\big(X^2+Y^2\big)$ and

$$q_2(Z) = q_2(X,Y) = \sigma\Big(\big(X^2+Y^2\big)^2 + 2\,X\big(XY - YX\big)Y\Big).$$

We propose the following problem.

Problem. Express the images $\Phi(q_j)$ explicitly for $j = 2,3,\cdots,n$.

We hope that the images $\Phi(q_j)$ for $j = 2,3,\cdots,n$ are expressed in the form of the *trace* as $\Phi(q_1)$.

Example 2.2.1. We consider the case $n = 1$. The algebra $\mathrm{Pol}(\mathbb{T}_1)^{U(1)}$ is generated by the polynomial

$$q(z) = z\,\overline{z}, \quad z \in \mathbb{C}.$$

Using Formula (2.2.26), we get

$$\Phi(q) = 4\,y^2\left(\frac{\partial^2}{\partial x^2} + \frac{\partial^2}{\partial y^2}\right).$$

Therefore $\mathbb{D}(\mathbb{H}_1) = \mathbb{C}\big[\Phi(q)\big]$.

Example 2.2.2. We consider the case $n = 2$. The algebra $\mathrm{Pol}(\mathbb{T}_2)^{U(2)}$ is generated by the polynomial

$$q_1(Z) = \sigma(Z\overline{Z}), \quad q_2(Z) = \sigma\Big(\big(Z\overline{Z}\big)^2\Big), \quad Z \in T_2.$$

Using Formula (2.2.26), we may express $\Phi(q_1)$ and $\Phi(q_2)$ explicitly. $\Phi(q_1)$ is expressed by Formula (2.2.28). The computation of $\Phi(q_2)$ might be quite tedious. We leave the detail to the reader. In this case, $\Phi(q_2)$ was essentially computed in [3], Proposition 6. Therefore $\mathbb{D}(\mathbb{H}_2) = \mathbb{C}\big[\Phi(q_1), \Phi(q_2)\big]$. The authors of [3] computed the center of the universal enveloping algebra $U(\mathfrak{sp}(2,\mathbb{C}))$ of $\mathfrak{sp}(2,\mathbb{C})$.

Now we describe the Siegel's fundamental domain for $\Gamma_n\backslash\mathbb{H}_n$. We let

$$\mathcal{P}_n = \Big\{ Y \in \mathbb{R}^{(n,n)} \mid Y = {}^tY > 0 \Big\}$$

be an open cone in $\mathbb{R}^d$ with $d = n(n+1)/2$. The general linear group $GL(n,\mathbb{R})$ acts on $\mathcal{P}_n$ transitively by

$$g \circ Y := gY\,{}^tg, \qquad g \in GL(n,\mathbb{R}),\ Y \in \mathcal{P}_n. \tag{2.2.29}$$

Thus $\mathcal{P}_n$ is a symmetric space diffeomorphic to $GL(n,\mathbb{R})/O(n)$.

The fundamental domain $\mathcal{R}_n$ for $GL(n,\mathbb{Z})\backslash\mathcal{P}_n$ which was found by H. Minkowski [25] is defined as a subset of $\mathcal{P}_n$ consisting of $Y = (y_{ij}) \in \mathcal{P}_n$ satisfying the following conditions (M.1)–(M.2) (cf. [14] p. 191 or [23] p. 123):

(M.1) $aY\,{}^ta \geq y_{kk}$ for every $a = (a_i) \in \mathbb{Z}^n$ in which $a_k, \cdots, a_n$ are relatively prime for $k = 1, 2, \cdots, n$.

(M.2) $y_{k,k+1} \geq 0$ for $k = 1, \cdots, n-1$.

We say that a point of $\mathcal{R}_n$ is Minkowski reduced or simply M-reduced. $\mathcal{R}_n$ has the following properties (R1)–(R4):

(R1) For any $Y \in \mathcal{P}_n$, there exist a matrix $A \in GL(n,\mathbb{Z})$ and $R \in \mathcal{R}_n$ such that $Y = R[A]$ (cf. [14] p. 191 or [23] p. 139). That is,

$$GL(n,\mathbb{Z}) \circ \mathcal{R}_n = \mathcal{P}_n.$$

(R2) $\mathcal{R}_n$ is a convex cone through the origin bounded by a finite number of hyperplanes. $\mathcal{R}_n$ is closed in $\mathcal{P}_n$ (cf. [23] p. 139).

(R3) If Y and $Y[A]$ lie in $\mathcal{R}_n$ for $A \in GL(n,\mathbb{Z})$ with $A \neq \pm I_n$, then Y lies on the boundary $\partial\mathcal{R}_n$ of $\mathcal{R}_n$. Moreover $\mathcal{R}_n \cap (\mathcal{R}_n[A]) \neq \emptyset$ for only finitely many $A \in GL(n,\mathbb{Z})$ (cf. [23] p. 139).

(R4) If $Y = (y_{ij})$ is an element of $\mathcal{R}_n$, then

$$y_{11} \le y_{22} \le \cdots \le y_{nn} \quad \text{and} \quad |y_{ij}| < \frac{1}{2} y_{ii} \quad \text{for } 1 \le i < j \le n.$$

We refer to [14] p. 192 or [23] pp. 123-124 for more detail.

Remark 6. Grenier [8] found another fundamental domain for $GL(n, \mathbb{Z})\backslash \mathcal{P}_n$.

For $Y = (y_{ij}) \in \mathcal{P}_n$, we put

$$dY = (dy_{ij}) \qquad \text{and} \qquad \frac{\partial}{\partial Y} = \left(\frac{1+\delta_{ij}}{2} \frac{\partial}{\partial y_{ij}} \right).$$

Then we can see easily that

$$ds^2 = \sigma((Y^{-1}dY)^2) \tag{2.2.30}$$

is a $GL(n, \mathbb{R})$-invariant Riemannian metric on $\mathcal{P}_n$ and its Laplacian is given by

$$\Delta = \sigma\left(\left(Y \frac{\partial}{\partial Y} \right)^2 \right).$$

We also can see that

$$d\mu_n(Y) = (\det Y)^{-\frac{n+1}{2}} \prod_{i \le j} dy_{ij}$$

is a $GL(n, \mathbb{R})$-invariant volume element on $\mathcal{P}_n$. The metric ds^2 on $\mathcal{P}_n$ induces the metric $ds^2_{\mathcal{R}}$ on $\mathcal{R}_n$. Minkowski [25] calculated the volume of $\mathcal{R}_n$ for the volume element $[dY] := \prod_{i \le j} dy_{ij}$ explicitly. Later Siegel computed the volume of $\mathcal{R}_n$ for the volume element $[dY]$ by a simple analytic method and generalized this case to the case of any algebraic number field.

Siegel [35] determined a fundamental domain $\mathcal{F}_n$ for $\Gamma_n \backslash \mathbb{H}_n$. We say that $\Omega = X + i\,Y \in \mathbb{H}_n$ with X, Y real is **Siegel reduced** or **S-reduced** if it has the following three properties:

(S.1) $\det(\mathrm{Im}\,(\gamma \cdot \Omega)) \le \det(\mathrm{Im}\,(\Omega))$ for all $\gamma \in \Gamma_n$;

(S.2) $Y = \mathrm{Im}\,\Omega$ is M-reduced, that is, $Y \in \mathcal{R}_n$;

(S.3) $|x_{ij}| \le \frac{1}{2}$ for $1 \le i, j \le n$, where $X = (x_{ij})$.

$\mathcal{F}_n$ is defined as the set of all Siegel reduced points in $\mathbb{H}_n$. Using the highest point method, Siegel proved the following (F1)–(F3) (cf. [14] pp. 194-197 or [23] p. 169):

(F1) $\Gamma_n \cdot \mathcal{F}_n = \mathbb{H}_n$, i.e., $\mathbb{H}_n = \cup_{\gamma \in \Gamma_n} \gamma \cdot \mathcal{F}_n$.

(F2) $\mathcal{F}_n$ is closed in $\mathbb{H}_n$.

(F3) $\mathcal{F}_n$ is connected and the boundary of $\mathcal{F}_n$ consists of a finite number of hyperplanes.

The metric ds^2 given by (2.2.1) induces a metric $ds^2_{\mathcal{F}}$ on $\mathcal{F}_n$. Siegel [35] computed the volume of $\mathcal{F}_n$

$$\mathrm{vol}\,(\mathcal{F}_n) = 2 \prod_{k=1}^{n} \pi^{-k}\, \Gamma(k)\, \zeta(2k), \tag{2.2.31}$$

where $\Gamma(s)$ denotes the Gamma function and $\zeta(s)$ denotes the Riemann zeta function. For instance,

$$\mathrm{vol}\,(\mathcal{F}_1) = \frac{\pi}{3}, \quad \mathrm{vol}\,(\mathcal{F}_2) = \frac{\pi^3}{270}, \quad \mathrm{vol}\,(\mathcal{F}_3) = \frac{\pi^6}{127575}, \quad \mathrm{vol}\,(\mathcal{F}_4) = \frac{\pi^{10}}{200930625}.$$

2.3 The Weil Representation

We recall that for a real symmetric positive definite matrix $c \in \mathbb{R}^{(m,m)}$, the Schrödinger representation U_c of $H_{\mathbb{R}}^{(n,m)}$ is defined by Formula (1.4.8). We refer to Formula (1.5.45). For convenience, we rewrite Formula (1.4.8)

$$(1.4.8) \qquad \big(U_c(g_0)f\big)(x) = e^{2\pi i\sigma(c(\kappa_0+\mu_0\,{}^t\lambda_0+2\,x\,{}^t\mu_0))}\,f(x+\lambda_0)$$

for $g_0 = (\lambda_0, \mu_0, \kappa_0) \in H_{\mathbb{R}}^{(n,m)},\ x \in \mathbb{R}^{(m,n)}$ and $f \in L^2\big(\mathbb{R}^{(m,n)}\big)$.

We let

$$G^J = Sp(n,\mathbb{R}) \ltimes H_{\mathbb{R}}^{(n,m)} \qquad \text{(semi-direct product)}$$

be the Jacobi group endowed with the following multiplication law

$$\Big(M,(\lambda,\mu,\kappa)\Big)\Big(M',(\lambda',\mu',\kappa')\Big) = \Big(MM',(\widetilde{\lambda}+\lambda',\widetilde{\mu}+\mu',\kappa+\kappa'+\widetilde{\lambda}\,{}^t\mu'-\widetilde{\mu}\,{}^t\lambda')\Big)$$

with $M, M' \in Sp(n,\mathbb{R}), (\lambda,\mu,\kappa),\ (\lambda',\mu',\kappa') \in H_{\mathbb{R}}^{(n,m)}$ and $(\widetilde{\lambda},\widetilde{\mu}) = (\lambda,\mu)M'$. Then $Sp(n,\mathbb{R})$ acts on $H_{\mathbb{R}}^{(n,m)}$ by conjugation inside G^J

$$M \star (\lambda,\mu,\kappa) = M(\lambda,\mu,\kappa)M^{-1} = (\lambda_*,\mu_*,\kappa), \tag{2.3.1}$$

where $M \in Sp(n,\mathbb{R}),\ (\lambda,\mu,\kappa) \in H_{\mathbb{R}}^{(n,m)}$ and $(\lambda_*,\mu_*) = (\lambda,\mu)M^{-1}$.

We fix an element $M \in Sp(n,\mathbb{R})$. We consider the mapping U_c^M of $H_{\mathbb{R}}^{(n,m)}$ into $\mathrm{Aut}\big(L^2\big(\mathbb{R}^{(m,n)}\big)\big)$ defined by

$$U_c^M(g) = U_c(M\star g) = U_c(MgM^{-1}), \qquad g \in H_{\mathbb{R}}^{(n,m)}. \tag{2.3.2}$$

Lemma 32. *U_c^M is an irreducible representation of $H_{\mathbb{R}}^{(n,m)}$ on $L^2\big(\mathbb{R}^{(m,n)}\big)$ such that*

$$U_c^M\big((0,0,\kappa)\big) = U_c\big((0,0,\kappa)\big) \qquad \text{for all } \kappa = {}^t\kappa \in \mathbb{R}^{(m,m)}.$$

Thus U_c^M is unitarily equivalent to U_c.

Proof. If $g_1, g_2 \in H_{\mathbb{R}}^{(n,m)}$, then

$$\begin{aligned} U_c^M(g_1g_2) &= U_c(M\star(g_1g_2)) = U_c\big(M(g_1g_2)M^{-1}\big) \\ &= U_c\big((Mg_1M^{-1})(Mg_2M^{-1})\big) \\ &= U_c\big(Mg_1M^{-1}\big)\,U_c\big(Mg_2M^{-1}\big) \\ &= U_c^M(g_1)\,U_c^M(g_2). \end{aligned}$$

Thus U_c^M is a representation of $H_{\mathbb{R}}^{(n,m)}$. The irreducibility of U_c^M follows immediately from that of U_c. It is easily seen that

$$U_c^M\big((0,0,\kappa)\big) = U_c\big(M\star(0,0,\kappa)\big) = U_c\big((0,0,\kappa)\big) \quad \text{for all } \kappa\in \mathrm{Sym}(n,\mathbb{R}).$$

Therefore it follows from Stone-von Neumann Theorem that U_c^M is unitarily equivalent to U_c. □

Since U_c^M is unitarily equivalent to U_c, there exists an unitary operator $R_c(M) : L^2\big(\mathbb{R}^{(m,n)}\big) \longrightarrow L^2\big(\mathbb{R}^{(m,n)}\big)$ such that $U_c^M(g)\,R_c(M) = R_c(M)\,U_c(g)$ for all $g\in H_{\mathbb{R}}^{(n,m)}$. For convenience, we take $R_c(I_{2n}) = I_c$, where I_c is the identity operator on $L^2\big(\mathbb{R}^{(m,n)}\big)$. We observe that $R_c(M)$ is determined uniquely up to a scalar of modulus one. For any two elements M_1, M_2 of $Sp(n,\mathbb{R})$, the unitary operator $R_c(M_2)^{-1}R_c(M_1)^{-1}R_c(M_1M_2)$ commutes with U_c. Indeed, for any element $g\in H_{\mathbb{R}}^{(n,m)}$, we have

$$\begin{aligned}
&U_c(g)\,R_c(M_2)^{-1}\,R_c(M_1)^{-1}\,R_c(M_1M_2)\\
&= R_c(M_2)^{-1}U_c^{M_2}(g)\,R_c(M_1)^{-1}\,R_c(M_1M_2)\\
&= R_c(M_2)^{-1}U_c(M_2\star g)\,R_c(M_1)^{-1}\,R_c(M_1M_2)\\
&= R_c(M_2)^{-1}\,R_c(M_1)^{-1}\,U_c^{M_1}(M_2\star g)\,R_c(M_1M_2)\\
&= R_c(M_2)^{-1}\,R_c(M_1)^{-1}\,U_c\big((M_1M_2)\star g\big)\,R_c(M_1M_2)\\
&= R_c(M_2)^{-1}\,R_c(M_1)^{-1}\,U_c^{M_1M_2}(g)\,R_c(M_1M_2)\\
&= R_c(M_2)^{-1}\,R_c(M_1)^{-1}\,R_c(M_1M_2)\,U_c(g).
\end{aligned}$$

According to Schur's lemma, we obtain a map $\alpha_c : Sp(n,\mathbb{R})\times Sp(n,\mathbb{R}) \longrightarrow \mathbb{C}_1^*$ satisfying the condition

$$R_c(M_1M_2) = \alpha_c(M_1,M_2)\,R_c(M_1)\,R_c(M_2) \tag{2.3.3}$$

for all $M_1, M_2\in Sp(n,\mathbb{R})$. Thus R_c is a projective representation of $Sp(n,\mathbb{R})$ with its multiplier α_c.

Lemma 33. *The map α_c satisfies the cocycle condition*

$$\alpha_c(M_1M_2,M_3)\,\alpha_c(M_1,M_2) = \alpha_c(M_1,M_2M_3)\,\alpha_c(M_2,M_3) \tag{2.3.4}$$

for all $M_1, M_2, M_3\in Sp(n,\mathbb{R})$.

Proof. Let $M_1, M_2, M_3\in Sp(n,\mathbb{R})$. Then according to Formula (2.3.3),

$$\begin{aligned}
R_c\big((M_1M_2)M_3\big) &= \alpha_c(M_1M_2,M_3)\,R_c(M_1M_2)\,R_c(M_3)\\
&= \alpha_c(M_1M_2,M_3)\,\alpha_c(M_1,M_2)\,R_c(M_1)\,R_c(M_2)\,R_c(M_3)
\end{aligned}$$

and

$$\begin{aligned}R_c\big(M_1(M_2M_3)\big) &= \alpha_c(M_1, M_2M_3)\, R_c(M_1)\, R_c(M_2M_3)\\ &= \alpha_c(M_1, M_2M_3)\, \alpha_c(M_2, M_3)\, R_c(M_1)\, R_c(M_2)\, R_c(M_3)\end{aligned}$$

Hence we obtain the cocycle condition (2.3.4). □

For $M = \begin{pmatrix} A & B \\ C & D \end{pmatrix} \in Sp(n, \mathbb{R})$ and $\Omega \in \mathbb{H}_n$, we put

$$J(M, \Omega) \;=\; \det(C\Omega + D) \tag{2.3.5}$$

and

$$J^*(M, \Omega) \;=\; \frac{J(M, \Omega)^{1/2}}{|J(M, \Omega)^{1/2}|}. \tag{2.3.6}$$

In fact, if $M_1, M_2 \in Sp(n, \mathbb{R})$, the cocycle $\alpha_c(M_1, M_2)$ is given by

$$\alpha_c(M_1, M_2) \;=\; \frac{J^*(M_1, iI_n)\, J^*(M_2, iI_n)}{J^*(M_1M_2, iI_n)}. \tag{2.3.7}$$

The cocycle α_c yields the central extension $Sp(n, \mathbb{R})_*$ of $Sp(n, \mathbb{R})$ by $\mathbb{C}_1^*$. The group $Sp(n, \mathbb{R})_*$ is the set $Sp(n, \mathbb{R}) \times \mathbb{C}_1^*$ with the following group multiplication

$$(M_1, t_1) \cdot (M_2, t_2) = \big(M_1M_2, t_1t_2\, \alpha_c(M_1, M_2)^{-1}\big) \tag{2.3.8}$$

for all $M_1, M_2 \in Sp(n, \mathbb{R})$ and $t_1, t_2 \in \mathbb{C}_1^*$. We see that the map $\widetilde{R}_c : Sp(n, \mathbb{R})_* \longrightarrow \mathrm{Aut}\big(L^2\big(\mathbb{R}^{(m,n)}\big)\big)$ defined by

$$\widetilde{R}_c(M, t) = t\, R_c(M), \qquad M \in Sp(n, \mathbb{R}),\ t \in \mathbb{C}_1^* \tag{2.3.9}$$

is a true representation of $Sp(n, \mathbb{R})_*$. We define the function $s_c : Sp(n, \mathbb{R}) \longrightarrow \mathbb{C}_1^*$ by

$$s_c(M) \;=\; |J(M, iI_n)|\, J(M, iI_n)^{-1}, \qquad M \in Sp(n, \mathbb{R}). \tag{2.3.10}$$

The following subset

$$Mp(n, \mathbb{R}) = \big\{(M, t) \in Sp(n, \mathbb{R})_* \mid t^2 = s_c(M)^{-1}\,\big\}$$

is a subgroup of $Sp(n, \mathbb{R})_*$ that is called the **metaplectic group**. We can show that $Mp(n, \mathbb{R})$ is a two-fold covering group of $Sp(n, \mathbb{R})$. The restriction ω_c of $\widetilde{R}_c$ to $Mp(n, \mathbb{R})$ is a true representation of $Mp(n, \mathbb{R})$ which is called the **Weil representation** of $Sp(n, \mathbb{R})$

Now we describe the action of ω_c explicitly. It is known that $Sp(n, \mathbb{R})$ is generated by the following elements

$$t_b = \begin{pmatrix} I_n & b \\ 0 & I_n \end{pmatrix} \quad \text{with } b = {}^t b \in \mathbb{R}^{(n,n)},$$
$$d_a = \begin{pmatrix} {}^t a & 0 \\ 0 & a^{-1} \end{pmatrix} \quad \text{with } a \in GL(n, \mathbb{R}),$$
$$\sigma_n = \begin{pmatrix} 0 & -I_n \\ I_n & 0 \end{pmatrix}.$$

Theorem 15. *The actions of R_c on the generators t_b, d_a and σ_n of $Sp(n, \mathbb{R})$ are given by*

$$(R_c(t_b)f)(x) = e^{2\pi i \sigma(c\, x\, b\, {}^t x)} f(x), \tag{2.3.11}$$
$$(R_c(d_a)f)(x) = (\det a)^{\frac{m}{2}} f(x\, {}^t a), \tag{2.3.12}$$
$$(R_c(\sigma_n)f)(x) = \left(\frac{2}{i}\right)^{\frac{mn}{2}} (\det c)^{\frac{n}{2}} \times \int_{\mathbb{R}^{(m,n)}} f(y) e^{-4\pi i\, \sigma(c\, y\, {}^t x)}\, dy, \tag{2.3.13}$$

where $f \in L^2\big(\mathbb{R}^{(m,n)}\big)$ and $x \in \mathbb{R}^{(m,n)}$.

Proof. Let $g = (\lambda, \mu, \kappa) \in H_{\mathbb{R}}^{(n,m)}$, $x \in \mathbb{R}^{(m,n)}$ and $f \in L^2\big(\mathbb{R}^{(m,n)}\big)$. For for each $t_b \in Sp(n, \mathbb{R})$ with $b = {}^t b \in \mathbb{R}^{(n,n)}$, we put

$$\big(T_c(t_b)f\big)(x) = e^{2\pi i \sigma(c\, x\, b\, {}^t x)} f(x) \qquad \text{for all } f \in L^2\big(\mathbb{R}^{(m,n)}\big).$$

Then

$$\begin{aligned}
\big(T_c(t_b)U_c(g)f\big)(x) &= e^{2\pi i \sigma(c\, x\, b\, {}^t x)} \big(U_c(g)f\big)(x) \\
&= e^{2\pi i \sigma(c\, x\, b\, {}^t x)} \cdot e^{2\pi i \sigma(c(\kappa + \mu\, {}^t\lambda + 2x\, {}^t\mu))} f(x+\lambda) \\
&= e^{2\pi i \sigma(c(\kappa + \mu\, {}^t\lambda + 2x\, {}^t\mu + x\, b\, {}^t x))} f(x+\lambda).
\end{aligned}$$

Since $t_b \star (\lambda, \mu, \kappa) = t_b(\lambda, \mu, \kappa)\, t_b^{-1} = (\lambda, -\lambda\, b + \mu, \kappa)$, we obtain

$$\begin{aligned}
&\big(U_c^{t_b}(g)\, T_c(t_b)f\big)(x) \\
&= \big(U_c(t_b \star g)\, T_c(t_b)f\big)(x) \\
&= \big(U_c(\lambda, -\lambda\, b + \mu, \kappa)\, T_c(t_b)f\big)(x) \\
&= e^{2\pi i \sigma(c(\kappa + (-\lambda\, b + \mu)\, {}^t\lambda + 2x\, {}^t(-\lambda b + \mu)))} \big(T_c(t_b)f\big)(x+\lambda) \\
&= e^{2\pi i \sigma(c(\kappa + (-\lambda\, b + \mu)\, {}^t\lambda + 2x\, {}^t(-\lambda b + \mu)))} \cdot e^{2\pi i \sigma(c(x+\lambda) b\, {}^t(x+\lambda))} f(x+\lambda). \\
&= e^{2\pi i \sigma(c(\kappa + \mu\, {}^t\lambda + 2x\, {}^t\mu + x\, b\, {}^t x))} f(x+\lambda).
\end{aligned}$$

Therefore

$$T_c(t_b)\,U_c(g)f = U_c^{t_b}(g)\,T_c(t_b)f$$

for all $b = {}^t b \in \mathbb{R}^{(n,n)}$, $g \in H_{\mathbb{R}}^{(n,m)}$ and $f \in L^2\big(\mathbb{R}^{(m,n)}\big)$.
Since $T_c(t_0) = T_c(I_{2n}) = I_c = R_c(I_{2n})$, we see that

$$R_c(t_b) = T_c(t_b) \qquad \text{for all } b = {}^t b \in \mathbb{R}^{(n,n)}.$$

We recall that I_c is the identity operator on $L^2\big(\mathbb{R}^{(m,n)}\big)$.

On the other hand, for each $f \in GL(n,\mathbb{R})$ and $f \in L^2\big(\mathbb{R}^{(m,n)}\big)$, we put

$$\big(A_c(d_a)f\big)(x) = (\det a)^{\frac{m}{2}}\, f(x\,{}^t a).$$

Then we have

$$\begin{aligned}
&\big(A_c(d_a)U_c(g)f\big)(x)\\
&= \big(\det a\big)^{\frac{m}{2}}\,\big(U_c(g)f\big)(x\,{}^t a)\\
&= \big(\det a\big)^{\frac{m}{2}}\, e^{2\pi i\,\sigma(c(\kappa+\mu\,{}^t\lambda+2\,x\,{}^t a\,{}^t\mu))} f(x\,{}^t a+\lambda).
\end{aligned}$$

Since $d_a \star (\lambda,\mu,\kappa) = d_a(\lambda,\mu,\kappa)\,d_a^{-1} = (\lambda\,{}^t a^{-1},\mu\,a,\kappa)$,

$$\begin{aligned}
&\big(U_c^{d_a}(g)\,A_c(d_a)f\big)(x)\\
&= \big(U_c(d_a\star g)\,A_c(d_a)f\big)(x)\\
&= \big(U_c(\lambda\,{}^t a^{-1},\mu\,a,\kappa)\,A_c(d_a)f\big)(x)\\
&= e^{2\pi i\,\sigma(c(\kappa+(\mu\,a)\,{}^t(\lambda\,{}^t a^{-1})+2\,x\,{}^t(\mu\,a)))}\,\big(A_c(d_a)f\big)\big(x+\lambda\,{}^t a^{-1}\big)\\
&= \big(\det a\big)^{\frac{m}{2}}\, e^{2\pi i\,\sigma(c(\kappa+\mu\,{}^t\lambda+2\,x\,{}^t a\,{}^t\mu))}\, f(x\,{}^t a+\lambda).
\end{aligned}$$

Thus

$$A_c(d_a)U_c(g)f = U_c^{d_a}(g)\,A_c(d_a)f$$

for all $a \in GL(n,\mathbb{R})$, $g \in H_{\mathbb{R}}^{(n,m)}$ and $f \in L^2\big(\mathbb{R}^{(m,n)}\big)$.
Since $A_c(d_{I_n}) = I_c = R_c(d_{I_n})$, we obtain $R_c(d_a) = A_c(d_a)$ for all $a \in GL(n,\mathbb{R})$.

Finally we put

$$\big(B_c(\sigma_n)f\big)(x) = \left(\frac{2}{i}\right)^{\frac{mn}{2}} (\det c)^{\frac{n}{2}} \int_{\mathbb{R}^{(m,n)}} f(y)\, e^{-4\pi i\,\sigma(c\,y\,{}^t x)}\, dy$$

for all $f \in L^2\big(\mathbb{R}^{(m,n)}\big)$.

$$
\begin{aligned}
&\big(B_c(\sigma_n)U_c(g)f\big)(x)\\
&= \left(\frac{2}{i}\right)^{\frac{mn}{2}} (\det c)^{\frac{n}{2}} \int_{\mathbb{R}^{(m,n)}} \big(U_c(g)f\big)(y)\, e^{-4\pi i\,\sigma(c\,y\,{}^t x)}\, dy\\
&= \left(\frac{2}{i}\right)^{\frac{mn}{2}} (\det c)^{\frac{n}{2}} \int_{\mathbb{R}^{(m,n)}} e^{2\pi i\,\sigma(c(\kappa+\mu\,{}^t\lambda+2\,y\,{}^t\mu))}\cdot e^{-4\pi i\,\sigma(c\,y\,{}^t x)}\, f(y+\lambda)\, dy\\
&= \left(\frac{2}{i}\right)^{\frac{mn}{2}} (\det c)^{\frac{n}{2}}\, e^{2\pi i\,\sigma(c(\kappa+\mu\,{}^t\lambda))} \int_{\mathbb{R}^{(m,n)}} e^{4\pi i\,\sigma(c\,y\,{}^t(\mu-x))}\, f(y+\lambda)\, dy\\
&= \left(\frac{2}{i}\right)^{\frac{mn}{2}} (\det c)^{\frac{n}{2}}\, e^{2\pi i\,\sigma(c(\kappa+\mu\,{}^t\lambda))} \int_{\mathbb{R}^{(m,n)}} e^{4\pi i\,\sigma(c\,(\widetilde{y}-\lambda)\,{}^t(\mu-x))}\, f(\widetilde{y})\, d\widetilde{y}\\
&= \left(\frac{2}{i}\right)^{\frac{mn}{2}} (\det c)^{\frac{n}{2}}\, e^{2\pi i\,\sigma(c(\kappa+\mu\,{}^t\lambda))}\cdot e^{-4\pi i\,\sigma(c\,\lambda\,{}^t(\mu-x))}\\
&\quad\times \int_{\mathbb{R}^{(m,n)}} f(y)\, e^{4\pi i\,\sigma(c\,y\,{}^t(\mu-x))}\, dy\\
&= \left(\frac{2}{i}\right)^{\frac{mn}{2}} (\det c)^{\frac{n}{2}}\, e^{2\pi i\,\sigma(c(\kappa-\lambda\,{}^t\mu+2\,x\,{}^t\lambda))} \int_{\mathbb{R}^{(m,n)}} f(y)\, e^{-4\pi i\,\sigma(c\,y\,{}^t(x-\mu))}\, dy.
\end{aligned}
$$

Since $\sigma_n \star (\lambda,\mu,\kappa) = \sigma_n(\lambda,\mu,\kappa)\,\sigma_n = (-\mu,\lambda,\kappa)$, we obtain

$$
\begin{aligned}
&\big(U_c^{\sigma_n}(g)\, B_c(\sigma_n)f\big)(x)\\
&= \big(U_c(\sigma_n\star g)\, B_c(\sigma_n)f\big)(x)\\
&= \big(U_c(-\mu,\lambda,\kappa)\, B_c(\sigma_n)f\big)(x)\\
&= e^{2\pi i\,\sigma(c(\kappa-\lambda\,{}^t\mu+2\,x\,{}^t\lambda))}\, \big(B_c(\sigma_n)f\big)(x-\mu)\\
&= \left(\frac{2}{i}\right)^{\frac{mn}{2}} (\det c)^{\frac{n}{2}}\, e^{2\pi i\,\sigma(c(\kappa-\lambda\,{}^t\mu+2\,x\,{}^t\lambda))} \int_{\mathbb{R}^{(m,n)}} f(y)\, e^{-4\pi i\,\sigma(c\,y\,{}^t(x-\mu))}\, dy.
\end{aligned}
$$

Therefore

$$B_c(\sigma_n)U_c(g)f = U_c^{\sigma_n}(g)\, B_c(\sigma_n)f \qquad \text{for all } f\in L^2\big(\mathbb{R}^{(m,n)}\big).$$

We note that we can take

$$R_c(\sigma_n) = B_c(\sigma_n).$$

Hence we complete the proof of the above theorem. □

Corollary 2. *We have the following*

(a) $\omega_c((t_b,1)) = R_c(t_b)$ and $\omega_c((t_b,-1)) = -R_c(t_b)$.

(b) If $\det a>0$, *then* $(d_a,\pm 1)\in Mp(n,\mathbb{R})$ *and hence we have*

$$\omega_c((d_a,1)) = R_c(d_a) \quad \text{and} \quad \omega_c((d_a,-1)) = -R_c(d_a).$$

(c) If $\det a < 0$, *then* $(d_a, \pm i) \in Mp(n,\mathbb{R})$ *and hence we have*

$$\omega_c((d_a,i)) = i\,R_c(d_a) \quad \text{and} \quad \omega_c((d_a,-i)) = -i\,R_c(d_a).$$

(d) $\omega_c\big((\sigma_n, i^{n/2})\big) = i^{n/2} R_c(\sigma_n)$ and $\omega_c\big((\sigma_n, -i^{n/2})\big) = -i^{n/2} R_c(\sigma_n)$.

Proof. The proof follows immediately from the definition of $Mp(n,\mathbb{R})$ and Theorem 15. □

Now we review some properties of ω_c. The Weil representation ω_c is not an irreducible representation of $Mp(n,\mathbb{R})$. In [15], Kashiwara and Vergne found an explicit decomposition of ω_c into irreducibles. First we observe that the orthogonal group $O(m)$ acts on $L^2\big(\mathbb{R}^{(m,n)}\big)$ by

$$(\alpha\cdot f)(x) \;=\; f(\alpha^{-1}x), \qquad \alpha\in O(m),\ x\in\mathbb{R}^{(m,n)},\ f\in L^2\big(\mathbb{R}^{(m,n)}\big).$$

This action commutes with ω_c. For each irreducible representation (σ, V_σ) of $O(m)$, we let $L^2\big(\mathbb{R}^{(m,n)};\sigma\big)$ be the space of all V_σ-valued square integrable functions $f:\mathbb{R}^{(m,n)} \longrightarrow V_\sigma$ satisfying the condition

$$f(\alpha^{-1}x) = \sigma(\alpha^{-1})f(x) \qquad \text{for all } \alpha\in O(m),\ x\in\mathbb{R}^{(m,n)}.$$

We let $\omega_c(\sigma)$ be the representation of $Mp(n,\mathbb{R})$ on $L^2\big(\mathbb{R}^{(m,n)};\sigma\big)$ defined by the formulas in Corollary 1. We denote by $\widehat{O(m)}$ the unitary dual of $O(m)$. In other words, $\widehat{O(m)}$ is the set of all equivalence classes of irreducible representations of $O(m)$. Let

$$\Sigma_m := \Big\{\, \sigma\in\widehat{O(m)} \;\Big|\; L^2\big(\mathbb{R}^{(m,n)};\sigma\big) \neq 0 \,\Big\}.$$

Kashiwara and Vergne proved that for any $\sigma\in\Sigma_m$, the representation $\omega_c(\sigma)$ is an irreducible unitary representation of $Mp(n,\mathbb{R})$ on $L^2\big(\mathbb{R}^{(m,n)};\sigma\big)$ and that ω_c is decomposed into irreducibles as follows:

$$\omega_c \;=\; \bigoplus_{\sigma\in\Sigma_m} (\dim V_\sigma)\,\omega_c(\sigma).$$

We realize $\omega_c(\sigma)$ in the space of vector valued holomorphic functions on $\mathbb{H}_n$. We note that $\mathbb{H}_n$ is biholomorphic to the Hermitian complex manifold $Sp(n,\mathbb{R})/K$ with $K := U(n)$ via the map

$$Sp(n,\mathbb{R})/K \longrightarrow \mathbb{H}_n, \qquad gK \mapsto g\cdot(iI_n),\quad M\in Sp(n,\mathbb{R}).$$

Let $\widehat{K}$ be the unitary dual of K. For any $(\tau, V_\tau)\in\widehat{K}$, we let $\mathcal{O}(\mathbb{H}_n, V_\tau)$ be the space of V_τ-valued holomorphic functions on $\mathbb{H}_n$. Let T_τ be the representation of $Mp(n,\mathbb{R})$ on $\mathcal{O}(\mathbb{H}_n, V_\tau)$ defined by

$$\big(T_\tau(M)f\big)(\Omega) := \tau({}^t(C\Omega + D))f(M^{-1}\cdot\Omega), \qquad (2.3.14)$$

where $M^{-1} = \begin{pmatrix} A & B \\ C & D \end{pmatrix} \in Sp(n,\mathbb{R}),\ f \in \mathcal{O}(\mathbb{H}_n, V_\tau)$ and $\Omega \in \mathbb{H}_n$. Here τ can be extended uniquely to a representation of the complexification $GL(n,\mathbb{C})$ of K. If v_τ is a highest weight vector of τ, then $\Phi_\tau(\Omega) := \tau(\Omega + iI_n)v_\tau$ is a a highest weight vector of T_τ. It can be shown that T_τ is an irreducible representation of $Sp(n,\mathbb{R})$ with highest weight vector Φ_τ.

Definition 9. A polynomial $f : \mathbb{R}^{(m,n)} \longrightarrow \mathbb{C}$ is called pluriharmonic if

$$\sum_{k=1}^{m} \frac{\partial^2 f}{\partial x_{ki}\partial x_{kj}} = 0 \qquad \text{for all } 1 \le i,j \le n.$$

Let $\mathfrak{H}$ be the space of all pluriharmonic polynomials on $\mathbb{R}^{(m,n)}$. Then $O(m)\times GL(n,\mathbb{R})$ acts on $\mathfrak{H}$ by

$$\big((\alpha,a)\cdot P\big) = P(\alpha^{-1}xa), \qquad \alpha \in O(m),\ a \in GL(n,\mathbb{R}),\ P \in \mathfrak{H}.$$

For $(\sigma, V_\sigma) \in \Sigma_m$, we let $\mathfrak{H}(\sigma)$ be the space of all V_σ-valued pluriharmonic polynomials $P : \mathbb{R}^{(m,n)} \longrightarrow V_\sigma$ such that

$$P(\alpha x) = \sigma(\alpha^{-1})^{-1}P(x) \qquad \text{for all } \alpha \in O(m) \text{ and } x \in \mathbb{R}^{(m,n)}.$$

Let $\tau(\sigma)$ be the representation of $GL(n,\mathbb{R})$ on $\mathfrak{H}(\sigma)$ defined by

$$\big(\tau(\sigma)(a)P\big)(x) = P(xa) \qquad a \in GL(n,\mathbb{R}),\ P \in \mathfrak{H}(\sigma).$$

For $\sigma \in \Sigma_m$, we see that $\mathfrak{H}(\sigma) \neq 0$ and $\tau(\sigma)$ is an irreducible finite dimensional representation of $GL(n,\mathbb{R})$ on $\mathfrak{H}(\sigma)$. They proved that the mapping $\sigma \longrightarrow \tau(\sigma)$ is an injection from Σ_m into $\widehat{GL(n,\mathbb{R})}$ and

$$\mathfrak{H} = \bigoplus_{\sigma\in\Sigma_m} \tau(\sigma)\otimes\sigma^* = \bigoplus_{\sigma\in\Sigma_m} \mathfrak{H}(\sigma)\otimes\sigma^*$$

as $O(m)\times GL(n,\mathbb{R})$-module.

Let $\sigma \in \Sigma_m$. We assume that $P : \mathbb{R}^{(m,n)} \longrightarrow \mathrm{Hom}_{\mathbb{C}}(V_{\tau(\sigma)}, V_\sigma)$ is a $\mathrm{Hom}_{\mathbb{C}}(V_{\tau(\sigma)}, V_\sigma)$-valued pluriharmonic polynomial on $\mathbb{R}^{(m,n)}$ satisfying the conditions

$$(A) \qquad P(\alpha x) = \sigma(\alpha^{-1})^{-1}P(x) \qquad \text{for all } \alpha \in O(m) \text{ and } x \in \mathbb{R}^{(m,n)}$$

and

$$(B) \qquad P(xa) = P(x)\,\big(\tau(\sigma)\otimes\det{}^{\frac{m}{2}}\big)(a) \qquad \text{for all } a \in GL(n,\mathbb{R}).$$

The unitary operator

$$\mathscr{F}_\sigma : L^2\big(\mathbb{R}^{(m,n)};\sigma\big) \longrightarrow \mathcal{O}\big(\mathbb{H}_n, V_{\tau(\sigma)}\big)$$

defined by

$$(\mathscr{F}_\sigma f)(\Omega) := \int_{\mathbb{R}^{(m,n)}} e^{\pi\, i\, \sigma(x\,\Omega\,{}^t x)}\, P(x)^* f(x)\, dx, \qquad f \in L^2\big(\mathbb{R}^{(m,n)};\sigma\big),\ \Omega \in \mathbb{H}_n$$

intertwines $\omega_c(\sigma)$ with $T_{\tau(\sigma)\otimes\det^{-\frac{m}{2}}}$.

2.4 Covariant Maps for the Weil Representation

Let c be a symmetric positive definite real matrix of degree m. We define the map $\mathscr{F}^{(c)} : \mathbb{H}_n \longrightarrow L^2\big(\mathbb{R}^{(m,n)}\big)$ by

$$\mathscr{F}^{(c)}(\Omega)(x) := e^{2\pi i \sigma(c x \Omega\, {}^t x)}, \qquad \Omega \in \mathbb{H}_n,\ x \in \mathbb{R}^{(m,n)}. \tag{2.4.1}$$

We define the automorphic factor $J_m : Sp(n,\mathbb{R}) \times \mathbb{H}_n \longrightarrow \mathbb{C}^*$ for $Sp(n,\mathbb{R})$ on $\mathbb{H}_n$ by

$$J_m(M,\Omega) = \det(C\Omega + D)^{\frac{m}{2}}, \tag{2.4.2}$$

where $M = \begin{pmatrix} A & B \\ C & D \end{pmatrix} \in Sp(n,\mathbb{R})$ and $\Omega \in \mathbb{H}_n$. We see that $Sp(n,\mathbb{R})$ acts on $\mathbb{H}_n$ transitively by

$$M \cdot \Omega = (A\Omega + B)(C\Omega + D)^{-1}, \tag{2.4.3}$$

where $M = \begin{pmatrix} A & B \\ C & D \end{pmatrix} \in Sp(n,\mathbb{R})$ and $\Omega \in \mathbb{H}_n$.

Theorem 16. *The map $\mathscr{F}^{(c)} : \mathbb{H}_n \longrightarrow L^2\big(\mathbb{R}^{(m,n)}\big)$ defined by Formula (2.4.1) is a covariant map for the Weil representation ω_c of $Mp(n,\mathbb{R})$ with respect to the automorphic factor J_m defined by Formula (2.4.2). In other words, $\mathscr{F}^{(c)}$ satisfies the following covariant relation*

$$R_c(M)\mathscr{F}^{(c)}(\Omega) = J_m(M,\Omega)^{-1}\, \mathscr{F}^{(c)}(M\cdot\Omega) \tag{2.4.4}$$

for all $M \in Sp(n,\mathbb{R})$ and $\Omega \in \mathbb{H}_n$. We recall that

$$\omega_c\big((M,t)\big) = t\, R_c(M) \qquad \text{(cf. (2.3.9))}$$

for all $(M,t) \in Mp(n,\mathbb{R})$ with $M \in Sp(n,\mathbb{R})$ and $t \in \mathbb{C}_1^*$.

Proof. For $M = \begin{pmatrix} A & B \\ C & D \end{pmatrix} \in Sp(n,\mathbb{R})$ and $\Omega \in \mathbb{H}_n$, we put

$$\Omega_* = M \cdot \Omega = (A\Omega + B)(C\Omega + D)^{-1}. \tag{2.4.5}$$

In this section, we use the notations t_b, d_a and σ_n in Section 2.3. It suffices to prove the covariance relation (2.4.4) for the generators $t_b\,(b = {}^t b \in \mathbb{R}^{(n,n)})$, $d_a\,(a \in GL(n,\mathbb{R}))$ and σ_n of $Sp(n,\mathbb{R})$.

Case I. $M = t_b$ with $b = {}^t b \in \mathbb{R}^{(n,n)}$.

In this case, we have

$$\Omega_* = \Omega + b.$$

By Formula (2.3.11) in Theorem 15,

$$\begin{aligned}&\left(R_c(M)\mathscr{F}^{(c)}(\Omega)\right)(x)\\=&\left(R_c(t_b)\mathscr{F}^{(c)}(\Omega)\right)(x)\\=&\,e^{2\pi i\sigma(c\,x\,b\,{}^t x)}\,\mathscr{F}^{(c)}(\Omega)(x).\end{aligned}$$

On the other hand, according to Formula (2.4.2),

$$\begin{aligned}&J_m(M,\Omega)^{-1}\mathscr{F}^{(c)}(M\cdot\Omega)(x)\\=&\,\mathscr{F}^{(c)}(\Omega+b)(x)\\=&\,e^{2\pi i\sigma(c\,x(\Omega+b)\,{}^t x))}\\=&\,e^{2\pi i\sigma(c\,x\,b\,{}^t x)}\,\mathscr{F}^{(c)}(\Omega)(x).\end{aligned}$$

Thus

$$R_c(t_b)\mathscr{F}^{(c)}(\Omega)=J_m(t_b,\Omega)^{-1}\mathscr{F}^{(c)}(t_b\cdot\Omega)$$

for all $b={}^t b\in\mathbb{R}^{(n,n)}$ and $\Omega\in\mathbb{H}_n$. Therefore we proved the covariance relation (2.4.4) in the case $M=t_b$ with $b={}^t b\in\mathbb{R}^{(n,n)}$.

Case II. $M=d_a=\begin{pmatrix}{}^t a & 0\\ 0 & a^{-1}\end{pmatrix}$ with $a\in GL(n,\mathbb{R})$.

In this case, we have

$$\Omega_*={}^t a\,\Omega\,a.$$

By Formula (2.3.12) in Theorem 15,

$$\begin{aligned}&\left(R_c(M)\mathscr{F}^{(c)}(\Omega)\right)(x)\\=&\,(\det a)^{\frac{m}{2}}\,\mathscr{F}^{(c)}(\Omega)(x\,{}^t a)\\=&\,(\det a)^{\frac{m}{2}}\,e^{2\pi i\sigma(c\,x\,{}^t a\,\Omega\,a\,{}^t x)}.\end{aligned}$$

On the other hand, according to Formula (2.4.2),

$$\begin{aligned}&J_m(M,\Omega)^{-1}\mathscr{F}^{(c)}(M\cdot\Omega)(x)\\=&\left(\det\left(a^{-1}\right)\right)^{-\frac{m}{2}}\,\mathscr{F}^{(c)}\left({}^t a\,\Omega\,a\right)(x)\\=&\,(\det a)^{\frac{m}{2}}\,e^{2\pi i\sigma(c\,x\,{}^t a\,\Omega\,a\,{}^t x)}.\end{aligned}$$

Thus

$$R_c(d_a)\mathscr{F}^{(c)}(\Omega)=J_m(d_a,\Omega)^{-1}\mathscr{F}^{(c)}(d_a\cdot\Omega)$$

for all $a\in GL(n,\mathbb{R})$ and $\Omega\in\mathbb{H}_n$. Therefore we proved the covariance relation (2.4.4) in the case $M=d_a$ with $d_a\in GL(n,\mathbb{R})$.

Case III. $M = \sigma_n = \begin{pmatrix} 0 & -I_n \\ I_n & 0 \end{pmatrix}$.

In this case, we have

$$\Omega_* = -\Omega^{-1} \qquad \text{and} \qquad J_m(M,\Omega) = (\det \Omega)^{\frac{m}{2}}.$$

In order to prove the covariance relation (2.4.4), we need the following useful lemma.

Lemma 34. *For a fixed element $\Omega \in \mathbb{H}_n$ and a fixed element $Z \in \mathbb{C}^{(m,n)}$, we obtain the following property*

$$\int_{\mathbb{R}^{(m,n)}} e^{\pi\, i\, \sigma(x\,\Omega\,{}^t x + 2\,x\,{}^t Z)} dx_{11} \cdots dx_{mn} \tag{2.4.6}$$
$$= \left(\det \frac{\Omega}{i}\right)^{-\frac{m}{2}} e^{-\pi\, i\, \sigma(Z\,\Omega^{-1}\,{}^t Z)},$$

where $x = (x_{ij}) \in \mathbb{R}^{(m,n)}$.

Proof of Lemma 34. By a simple computation, we see that

$$e^{\pi i\,\sigma(x\Omega\,{}^t x + 2x\,{}^t Z)} = e^{-\pi i\,\sigma(Z\Omega^{-1}\,{}^t Z)} \cdot e^{\pi i\,\sigma\{(x+Z\Omega^{-1})\Omega\,{}^t(x+Z\Omega^{-1})\}}.$$

We observe that the real Jacobi group $Sp(n,\mathbb{R}) \ltimes H_{\mathbb{R}}^{(n,m)}$ acts on $\mathbb{H}_n \times \mathbb{C}^{(m,n)}$ holomorphically and transitively by

$$\big(M,(\lambda,\mu,\kappa)\big)\cdot(\Omega,Z) = \big(M\cdot\Omega, (Z+\lambda\,\Omega+\mu)(C\Omega+D)^{-1}\big), \tag{2.4.7}$$

where $M \in Sp(n,\mathbb{R})$, $(\lambda,\mu,\kappa) \in H_{\mathbb{R}}^{(n,m)}$, $\Omega \in \mathbb{H}_n$ and $Z \in \mathbb{C}^{(m,n)}$. So we may put

$$\Omega = i\,A\,{}^t A, \quad Z = iV, \quad A \in \mathbb{R}^{(n,n)}, \quad V = (v_{ij}) \in \mathbb{R}^{(m,n)}.$$

Then we obtain

$$\int_{\mathbb{R}^{(m,n)}} e^{\pi i\,\sigma(x\Omega\,{}^t x + 2x\,{}^t Z)} dx_{11} \cdots dx_{mn}$$
$$= e^{-\pi i\,\sigma(Z\Omega^{-1}\,{}^t Z)}$$
$$\times \int_{\mathbb{R}^{(m,n)}} e^{\pi i\,\sigma[\{x+iV(iA\,{}^t A)^{-1}\}(iA\,{}^t A)\,{}^t\{x+iV(iA\,{}^t A)^{-1}\}]} dx_{11} \cdots dx_{mn}$$

$$= e^{-\pi i\,\sigma(Z\Omega^{-1}\,{}^tZ)} \int_{\mathbb{R}^{(m,n)}} e^{\pi i\,\sigma[\{x+V(A\,{}^tA)^{-1}\}A\,{}^tA\,{}^t\{x+V(A\,{}^tA)^{-1}\}]}\,dx_{11}\cdots dx_{mn}$$

$$= e^{-\pi i\,\sigma(Z\Omega^{-1}\,{}^tZ)} \int_{\mathbb{R}^{(m,n)}} e^{-\pi\,\sigma\{(uA)\,{}^t(uA)\}}\,du_{11}\cdots du_{mn}$$

$$\big(\text{Put } u = x + V(A\,{}^tA)^{-1} = (u_{ij})\big)$$

$$= e^{-\pi i\,\sigma(Z\Omega^{-1}\,{}^tZ)} \int_{\mathbb{R}^{(m,n)}} e^{-\pi\,\sigma(w\,{}^tw)}(\det A)^{-m}\,dw_{11}\cdots dw_{mn}$$

$$\big(\text{Put } w = uA = (w_{ij})\big)$$

$$= e^{-\pi i\,\sigma(Z\Omega^{-1}\,{}^tZ)}\,(\det A)^{-m}\cdot\left(\prod_{i=1}^{m}\prod_{j=1}^{n}\int_{\mathbb{R}} e^{-\pi\,w_{ij}^2}\,dw_{ij}\right)$$

$$= e^{-\pi i\,\sigma(Z\Omega^{-1}\,{}^tZ)}\,(\det A)^{-m}\quad\Big(\text{because } \int_{\mathbb{R}} e^{-\pi\,w_{ij}^2}\,dw_{ij} = 1\quad\text{for all } i,j\Big)$$

$$= e^{-\pi i\,\sigma(Z\Omega^{-1}\,{}^tZ)}\,\big(\det\,(A\,{}^tA)\big)^{-\frac{m}{2}}$$

$$= e^{-\pi i\,\sigma(Z\Omega^{-1}\,{}^tZ)}\left(\det\left(\frac{\Omega}{i}\right)\right)^{-\frac{m}{2}}.$$

This completes the proof of Lemma 34. □

According to Formula (2.3.13) in Theorem 15,

$$\begin{aligned}
&\Big(R_c(\sigma_n)\mathscr{F}^{(c)}(\Omega)\Big)(x)\\
&= \left(\frac{2}{i}\right)^{\frac{mn}{2}}(\det c)^{\frac{n}{2}}\int_{\mathbb{R}^{(m,n)}}\mathscr{F}^{(c)}(\Omega)(y)\,e^{-4\pi i\,\sigma(c\,y\,{}^tx)}\,dy\\
&= \left(\frac{2}{i}\right)^{\frac{mn}{2}}(\det c)^{\frac{n}{2}}\int_{\mathbb{R}^{(m,n)}} e^{2\pi i\,\sigma(c\,y\,\Omega\,{}^ty)}\cdot e^{-4\pi i\,\sigma(c\,y\,{}^tx)}\,dy\\
&= \left(\frac{2}{i}\right)^{\frac{mn}{2}}(\det c)^{\frac{n}{2}}\int_{\mathbb{R}^{(m,n)}} e^{\pi i\,\sigma\{c\,(y\,(2\,\Omega)\,{}^ty\,+\,2\,y\,{}^t(-2\,x))\}}\,dy
\end{aligned}$$

If we substitute $u = c^{1/2}\,y$, then $du = \big(\det c\big)^{\frac{n}{2}}\,dy$. Therefore according to Lemma 34, we obtain

$$\begin{aligned}
&\Big(R_c(\sigma_n)\mathscr{F}^{(c)}(\Omega)\Big)(x)\\
&=\left(\frac{2}{i}\right)^{\frac{mn}{2}}(\det c)^{\frac{n}{2}}\int_{\mathbb{R}^{(m,n)}} e^{\pi\, i\,\sigma(u\,(2\,\Omega)\,{}^t u+2\,c^{1/2}\,u\,{}^t(-2\,x))}\,(\det c)^{-\frac{n}{2}}\,du\\
&=\left(\frac{2}{i}\right)^{\frac{mn}{2}}\int_{\mathbb{R}^{(m,n)}} e^{\pi\, i\,\sigma(u\,(2\,\Omega)\,{}^t u+2\,u\,{}^t(-2\,c^{1/2}\,x))}\,du\\
&=\left(\frac{2}{i}\right)^{\frac{mn}{2}}\left(\det\frac{2\,\Omega}{i}\right)^{-\frac{m}{2}} e^{-\pi\, i\,\sigma((-2\,c^{1/2}\,x)\,(2\,\Omega)^{-1}\,{}^t(-2\,c^{1/2}\,x))}\\
&=\big(\det\Omega\big)^{-\frac{m}{2}}\,e^{-2\,\pi\, i\,\sigma(c\,x\,\Omega^{-1}\,{}^t x)}.
\end{aligned}$$

On the other hand, according to Formula (2.4.2),

$$\begin{aligned}
&J_m(M,\Omega)^{-1}\mathscr{F}^{(c)}(M\cdot\Omega)(x)\\
&=J_m(\sigma,\Omega)^{-1}\,\mathscr{F}^{(c)}\big(-\Omega^{-1}\big)(x)\\
&=\big(\det\Omega\big)^{-\frac{m}{2}}\,e^{2\,\pi\, i\,\sigma(c\,x\,(-\Omega^{-1})\,{}^t x)}\\
&=\big(\det\Omega\big)^{-\frac{m}{2}}\,e^{-2\,\pi\, i\,\sigma(c\,x\,\Omega^{-1}\,{}^t x)}.
\end{aligned}$$

So we see that

$$R_c(\sigma_n)\mathscr{F}^{(c)}(\Omega)=J_m(\sigma_n,\Omega)^{-1}\,\mathscr{F}^{(c)}(\sigma_n\cdot\Omega). \tag{2.4.8}$$

Therefore the covariance relation (2.4.4) holds for the case $\sigma_n=\begin{pmatrix}0&-I_n\\ I_n&0\end{pmatrix}$.

Since J_m is an automorphic factor for $Sp(n,\mathbb{R})$ on $\mathbb{H}_n$, we see that if the covariance relation (2.4.4) holds for M_1, M_2 in $Sp(n,\mathbb{R})$, then it holds for M_1M_2. Finally we complete the proof. □

Now we can give another realization of the metaplectic group $Mp(n,\mathbb{R})$ that was dealt with in Section 2.1 and Section 2.3.

Proposition 9. *Let $(U_c,\mathcal{H}_c)$ be the Schrödinger representation of the Heisenberg group $H_{\mathbb{R}}^{(n,m)}$ defined by Formula (1.4.8) with the model $\mathcal{H}_c=L^2\big(\mathbb{R}^{(m,n)},d\xi\big)$. We denote by $U(\mathcal{H}_c)$ the group of all unitary isomorphisms of $\mathcal{H}_c$. Let $\widetilde{Mp(c)}$ be the set of all $R\in U(\mathcal{H}_c)$ such that*

$$R\,U_c(g)=U_c(M\star g)R=U_c(MgM^{-1})R$$

for all $g\in H_{\mathbb{R}}^{(n,m)}$ and for some $M\in Sp(n,\mathbb{R})$. Then for a given element $R\in\widetilde{Mp(c)}$, the corresponding $M\in Sp(n,\mathbb{R})$ is determined uniquely, denoted by $M=\nu_c(R)$. Moreover there is an exact sequence of groups

$$1 \longrightarrow \mathbb{C}_1^* \longrightarrow \widetilde{Mp(c)} \xrightarrow{\nu_c} Sp(n,\mathbb{R}) \longrightarrow 1. \tag{2.4.9}$$

Proof. First of all we observe that $\widetilde{Mp(c)}$ is a subgroup of $U(\mathcal{H}_c)$. Let $R \in \widetilde{Mp(c)}$, and $M_1, M_2 \in Sp(n,\mathbb{R})$ such that

$$R\,U_c(g) = U_c(M_1 \star g)\,R = U_c(M_2 \star g)\,R \qquad \text{for all } g \in H_{\mathbb{R}}^{(n,m)}.$$

Then $U_c(M_1 \star g) = U_c(M_2 \star g)$ for all $g \in H_{\mathbb{R}}^{(n,m)}$. According to Formula (1.4.8), $(M_1^{-1}M_2)g = g(M_1^{-1}M_2)$ for all $g \in H_{\mathbb{R}}^{(n,m)}$. Thus $M_1 = M_2$. It follows that the map $\nu_c : \widetilde{Mp(c)} \longrightarrow Sp(n,\mathbb{R})$ is well defined. it is easily checked that ν_c is a group homomorphism. The kernel of ν_c is given by

$$\ker \nu_c = \left\{ R \in U(\mathcal{H}_c) \,\middle|\, R\,U_c(g) = U_c(g)\,R \quad \text{for all } g \in H_{\mathbb{R}}^{(n,m)} \right\}.$$

Since U_c is irreducible and unitary, according to Schur's lemma, $\ker \nu_c = \mathbb{C}_1^*$. The surjectivity of ν_c follows from the arguments in Section 2.3. □

According to Theorem 15, $R_c(t_b)$, $R_c(d_a)$ and $R_c(\sigma_n)$ are members of $\widetilde{Mp(c)}$ sitting above the generators t_b, d_a and σ_n of $Sp(n,\mathbb{R})$ respectively. That is, $\nu_c(R_c(t_b)) = t_b$, $\nu_c(R_c(d_a)) = d_a$ and $\nu_c(R_c(\sigma_n)) = \sigma_n$.

Theorem 17. *Let $P \in \widetilde{Mp(c)}$ and $\nu_c(P) = M = \begin{pmatrix} A & B \\ C & D \end{pmatrix} \in Sp(n,\mathbb{R})$. Then for any $\Omega \in \mathbb{H}_n$,*

$$P\mathscr{F}^{(c)}(\Omega) = B_c(P;\Omega)\mathscr{F}^{(c)}(M \cdot \Omega),$$

where $B_c(P;\Omega)$ is, up to a scalar of absolute one, a branch of the holomorphic function $\left\{ \det(C\Omega + D)^{\frac{1}{2}} \right\}^{-m}$ on $\mathbb{H}_n$.

Proof. Let $\mathbb{G}_1$ be the subgroup of $\widetilde{Mp(c)}$ consisting of all $P \in \widetilde{Mp(c)}$ such that

$$P\mathscr{F}^{(c)}(\Omega) = c_P\,\mathscr{F}^{(c)}\big(\nu_c(P)\cdot\Omega\big) \quad \text{for all } \Omega \in \mathbb{H}_n,$$

where c_P is a constant depending only on P. For $P \in \mathbb{G}_1$, we write

$$P\mathscr{F}^{(c)}(\Omega) = B_c(P;\Omega)\,\mathscr{F}^{(c)}\big(\nu_c(P)\cdot\Omega\big) \quad \text{for all } \Omega \in \mathbb{H}_n.$$

Let $\mathbb{G}_2$ be the set of all $P \in \mathbb{G}_1$ satisfying the following conditions ($\mathbb{G}$1) and ($\mathbb{G}$2):

($\mathbb{G}$1) $B_c(P;\Omega)$ is continuous in $\Omega \in \mathbb{H}_n$;

($\mathbb{G}$2) $\{B_c(P;\Omega)\}^2\,|\det(C\Omega + D)|^m$ is independent of Ω with values in $\mathbb{C}_1^*$ for $\nu_c(P) = \begin{pmatrix} A & B \\ C & D \end{pmatrix} \in Sp(n,\mathbb{R})$.

It is easily checked that for $P, Q \in \mathbb{G}_2$,

$$B_c(QP;\Omega) = B_c(P;\Omega)\, B_c(Q;\nu_2(P)\cdot\Omega) \qquad \text{for all } \Omega\in\mathbb{H}_n. \tag{2.4.10}$$

Indeed, we get

$$\begin{aligned}(QP)\mathscr{F}^{(c)}(\Omega) &= Q\big(P\mathscr{F}^{(c)}(\Omega)\big)\\ &= B_c(P;\Omega)\Big(Q\big(\mathscr{F}^{(c)}(\nu_c(P)\cdot\Omega)\big)\Big)\\ &= B_c(P;\Omega)\, B_c(Q;\nu_c(P)\cdot\Omega)\,\mathscr{F}^{(c)}\big(\nu_c(Q)\cdot(\nu_c(P)\cdot\Omega)\big)\\ &= B_c(P;\Omega)\, B_c(Q;\nu_c(P)\cdot\Omega)\,\mathscr{F}^{(c)}\big(\nu_c(QP)\cdot\Omega\big).\end{aligned}$$

By Formula (2.4.9) together with the fact that $J(M,\Omega) := \det(C\Omega + D)$ for $M = \begin{pmatrix} A & B\\ C & D\end{pmatrix} \in Sp(n,\mathbb{R})$ and $\Omega\in\mathbb{H}_n$ is automorphic factor, we see that $\mathbb{G}_2$ is a subgroup of $\mathbb{G}_1$. We observe that $R_c(t_b)$, $R_c(d_a)$, $R_c(\sigma_n)$ in Theorem 15 and $\alpha\in\mathbb{C}_1^*$ generate the group $\widetilde{Mp(c)}$. We shall show that $R_c(t_b)$, $R_c(d_a)$, $R_c(\sigma_n)$ and $\alpha\in\mathbb{C}_1^*$ belong to $\mathbb{G}_2$. Then $\mathbb{G}_1 = \mathbb{G}_2 = \widetilde{Mp(c)}$. This implies the proof of the theorem.

Now we shall prove that $R_c(t_b)$, $R_c(d_a)$, $R_c(\sigma_n)$ and $\alpha\in\mathbb{C}_1^*$ belong to $\mathbb{G}_2$. For brevity we put $F_c(P;\Omega) = \{B_c(P;\Omega)\}^2\,|\det(C\Omega+D)|^m$ for $\nu_c(P) = \begin{pmatrix} A & B\\ C & D\end{pmatrix}\in Sp(n,\mathbb{R})$ with $P\in\widetilde{Mp(c)}$.

Case I. $P = \alpha\in\mathbb{C}_1^*\subset\widetilde{Mp(c)}$.

In this case, we obtain

$$P\mathscr{F}^{(c)}(\Omega) = \alpha\,\mathscr{F}^{(c)}(\Omega).$$

So we get $B_c(P;\Omega) = \alpha$ and $F_c(P;\Omega) = \alpha^2$. Thus $\alpha\in\mathbb{G}_2$.

Case II. $P = R_c(t_b)$ with $t_b = \begin{pmatrix} I_n & b\\ 0 & I_n\end{pmatrix}\in Sp(n,\mathbb{R})$.

In this case, according to Formula (2.3.11), we obtain

$$\begin{aligned}P\mathscr{F}^{(c)}(\Omega) &= e^{2\pi i\,\sigma(c\,xb\,{}^tx)}\,\mathscr{F}^{(c)}(\Omega)(x)\\ &= e^{2\pi i\,\sigma\{c\,x(\Omega+b)\,{}^tx\}}\\ &= \mathscr{F}^{(c)}(\Omega+b)(x) = \mathscr{F}^{(c)}(t_b\cdot\Omega)(x)\\ &= \mathscr{F}^{(c)}\big(\nu_c(R_c(t_b))\cdot\Omega\big)(x).\end{aligned}$$

We get $B_c(P;\Omega) = 1$ and $F_c(P;\Omega) = 1$. Thus $R_c(t_b)\in\mathbb{G}_2$.

Case III. $P = R_c(d_a)$ with $d_a = \begin{pmatrix} {}^ta & 0\\ 0 & a^{-1}\end{pmatrix}\in Sp(n,\mathbb{R})$.

In this case, according to Formula (2.3.12), we obtain

$$\begin{aligned} P\mathscr{F}^{(c)}(\Omega) &= (\det a)^{\frac{m}{2}}\,\mathscr{F}^{(c)}(\Omega)(x\,{}^t a) \\ &= (\det a)^{\frac{m}{2}}\,e^{2\pi i\,\sigma\{c\,x({}^t a\,\Omega\,a)\,{}^t x\}} \\ &= (\det a)^{\frac{m}{2}}\,\mathscr{F}^{(c)}(d_a\cdot\Omega)(x) \\ &= (\det a)^{\frac{m}{2}}\,\mathscr{F}^{(c)}\big(\nu_c(R_c(d_a))\cdot\Omega\big)(x). \end{aligned}$$

We get $B_c(P;\Omega) = (\det a)^{\frac{m}{2}}$ and $F_c(P;\Omega) = 1$. Thus $R_c(d_a) \in \mathbb{G}_2$.

Case IV. $P = R_c(\sigma_n)$ with $\sigma_n = \begin{pmatrix} 0 & -I_n \\ I_n & 0 \end{pmatrix} \in Sp(n,\mathbb{R})$.

In this case, according to Formula (2.3.13), we obtain

$$\begin{aligned} P\mathscr{F}^{(c)}(\Omega) &= \left(\frac{2}{i}\right)^{\frac{mn}{2}} (\det c)^{\frac{n}{2}} \int_{\mathbb{R}^{(m,n)}} \mathscr{F}^{(c)}(\Omega)(y)\,e^{-4\pi i\,\sigma(c\,y\,{}^t x)}\,dy \\ &= \left(\frac{2}{i}\right)^{\frac{mn}{2}} (\det c)^{\frac{n}{2}} \int_{\mathbb{R}^{(m,n)}} e^{2\pi i\,\sigma\{c\,(y\,\Omega\,{}^t y - 2\,y{}^t x)\}}\,dy \\ &= (\det\Omega)^{-\frac{m}{2}}\,e^{-2\pi i\,\sigma(c\,x\,\Omega^{-1}\,{}^t x)} \qquad \text{(by Lemma 14.2)} \\ &= (\det\Omega)^{-\frac{m}{2}}\,\mathscr{F}^{(c)}(-\Omega^{-1})(x) \\ &= (\det\Omega)^{-\frac{m}{2}}\,\mathscr{F}^{(c)}\big(\nu_c(R_c(\sigma_n))\cdot\Omega\big)(x). \end{aligned}$$

We get $B_c(P;\Omega) = (\det\Omega)^{-\frac{m}{2}}$ with $B_c(P;iI_n) = i^{-\frac{mn}{2}}$, and $F_c(P;\Omega) = i^{-\frac{mn}{2}}$. Thus $R_c(\sigma_n) \in \mathbb{G}_2$. Hence we complete the proof. □

Definition 10. Let $\chi_c : \widetilde{Mp(c)} \longrightarrow \mathbb{C}$ be the map defined by

$$\chi_c(P) = \det(C\Omega + D)^m \{B_c(P;\Omega)\}^2, \quad P \in \widetilde{Mp(c)},$$

where $\nu_c(P) = \begin{pmatrix} A & B \\ C & D \end{pmatrix} \in Sp(n,\mathbb{R})$. According to Theorem 17, the image of χ_c is contained in $\mathbb{C}_1^*$ and $\chi_c : \widetilde{Mp(c)} \longrightarrow \mathbb{C}_1^*$ is a character of $\widetilde{Mp(c)}$. Furthermore we have

$$\chi_c(\alpha) = \alpha^2 \quad \text{for any } \alpha \in \mathbb{C}_1^* \subset \widetilde{Mp(c)}.$$

We denote by $Mp(n,\mathbb{R})_c$ the kernel of χ_c. We call $Mp(n,\mathbb{R})_c$ the *metaplectic group* attached to U_c.

We let

$$m_\diamond : \widetilde{Mp(c)} \times \widetilde{Mp(c)} \longrightarrow \widetilde{Mp(c)}$$

be the multiplication map and let

$$\Phi_{[c]} : \widetilde{Mp(c)} \times \mathbb{H}_n \longrightarrow \mathbb{C}^*$$

be the map defined by

$$\Phi_{[c]}(P,\Omega) := B_c(P;\Omega), \quad P \in \widetilde{Mp(c)},\ \Omega \in \mathbb{H}_n.$$

We provide $\widetilde{Mp(c)}$ with the weakest topology such that the following three maps

$$\nu_c : \widetilde{Mp(c)} \longrightarrow Sp(n,\mathbb{R}), \quad m_\diamond : \widetilde{Mp(c)} \times \widetilde{Mp(c)} \longrightarrow \widetilde{Mp(c)},$$
$$\Phi_{[c]} : \widetilde{Mp(c)} \times \mathbb{H}_n \longrightarrow \mathbb{C}^*$$

are all continuous.

Then we have the following properties.

Lemma 35. *$\widetilde{Mp(c)}$ is a Hausdorff space on the above weakest topology.*

Proof. Fix an element $\Omega_0 \in \mathbb{H}_n$. Let $\eta : \widetilde{Mp(c)} \longrightarrow Sp(n,\mathbb{R}) \times \mathbb{C}^*$ by

$$\eta(P) := \big(\nu_c(P), B_c(P;\Omega_0)\big), \quad P \in \widetilde{Mp(c)}. \tag{2.4.11}$$

Then by the weak topology on $\widetilde{Mp(c)}$, η is continuous. If $P, Q \in \widetilde{Mp(c)}$ such that $\eta(P) = \eta(Q)$, then $\nu_c(P) = \nu_c(Q)$ and $B_c(P;\Omega_0) = B_c(Q;\Omega_0)$. $QP^{-1} = \alpha \in \mathbb{C}_1^*$ because $\nu_c(QP^{-1}) = 1$. Thus $Q = \alpha P$. By assumption,

$$B_c(Q;\Omega_0) = B_c(\alpha P;\Omega_0) = \alpha B_c(P;\Omega_0) = B_c(P;\Omega_0).$$

Therefore $\alpha = 1$, that is, $P = Q$. This implies that η is one-to-one.

Let $f : Sp(n,\mathbb{R}) \times \mathbb{C}^* \longrightarrow \mathbb{C}^*$ be the map defined by

$$f(M,\alpha) := \alpha^2 \left\{\det(C\Omega_0 + D)\right\}^m, \tag{2.4.12}$$

where $M = \begin{pmatrix} A & B \\ C & D \end{pmatrix} \in Sp(n,\mathbb{R})$ and $\alpha \in \mathbb{C}^*$. By Theorem 17, $\eta\big(\widetilde{Mp(c)}\big) = f^{-1}(\mathbb{C}_1^*)$. Since $\frac{\partial f}{\partial \alpha} \neq 0$ and $\mathbb{C}_1^*$ is a submanifold of $\mathbb{C}^*$, we see that $\eta\big(\widetilde{Mp(c)}\big)$ is a submanifold of $Sp(n,\mathbb{R}) \times \mathbb{C}^*$. Therefore $\eta\big(\widetilde{Mp(c)}\big)$ is Hausdorff because $Sp(n,\mathbb{R})$ and $\mathbb{C}^*$ are Hausdorff. □

Lemma 36. *Let $h : \widetilde{Mp(c)} \longrightarrow Sp(n,\mathbb{R}) \times \mathbb{C}_1^*$ be the map defined by*

$$h(P) := \big(\nu_c(P), \chi_c(P)\big), \quad P \in \widetilde{Mp(c)}.$$

Then the map h defines a connected double covering of the Lie group $Sp(n,\mathbb{R}) \times \mathbb{C}_1^*$, and hence gives $\widetilde{Mp(c)}$ the structure of a Lie group.

Proof. We note that h is continuous. We see that

$$\ker h = \ker \nu_c \cap \ker \chi_c = \mathbb{C}_1^* \cap \ker \chi_c = \ker \left(\chi_c|_{\mathbb{C}_1^*}\right) = \{\pm 1\}.$$

Let $h_* : Sp(n,\mathbb{R}) \times \mathbb{C}^* \longrightarrow Sp(n,\mathbb{R}) \times \mathbb{C}^*$ be the map defined by

$$h_*(M,\alpha) := (M, f(M,\alpha)), \quad M \in Sp(n,\mathbb{R}),\ \alpha \in \mathbb{C}^*, \tag{2.4.13}$$

where f is the map defined by (2.4.12). Then $h = h_* \circ \eta$, where η is the map defined by (2.4.11). Clearly h_* is a double covering projection. Since $h_*^{-1}\left(Sp(n,\mathbb{R} \times \mathbb{C}_1^*\right) = \eta\big(\widetilde{Mp(c)}\big)$, the restriction $h_{*,\eta}$ of h_* to $\eta\big(\widetilde{Mp(c)}\big)$ is a double covering

$$h_{*,\eta} : \widetilde{Mp(c)} \longrightarrow Sp(n,\mathbb{R}) \times \mathbb{C}_1^*$$

of the manifold $Sp(n,\mathbb{R}) \times \mathbb{C}_1^*$. It only remains to prove that $\widetilde{Mp(c)}$ is connected. Since $Sp(n,\mathbb{R})$ and $\mathbb{C}_1^*$ are connected, according to the exact sequence (2.4.9), $\widetilde{Mp(c)}$ is connected. □

Proposition 10. *$Mp(n,\mathbb{R})_c$ is a closed connected subgroup of $\widetilde{Mp(c)}$ and $q_c : Mp(n,\mathbb{R})_c \longrightarrow Sp(n,\mathbb{R})$ is a double covering projection with $\ker q_c = \{\pm 1\}$, where q_c is the restriction of ν_c to $Mp(n,\mathbb{R})_c$.*

Proof. By Lemma 36, q_c is a double covering projection of $Sp(n,\mathbb{R})$. Ir only remains to prove that $Mp(n,\mathbb{R})_c$ is connected. The stabilizer at $i\,I_n$ under the action (2.1.6) of $Sp(n,\mathbb{R})$ is given by

$$\left\{ \begin{pmatrix} A & B \\ -B & A \end{pmatrix} \in Sp(n,\mathbb{R}) \;\middle|\; A\,{}^tA + B\,{}^tB = I_n,\ A\,{}^tB = B\,{}^tB \right\}$$

that is isomorphic to $U(n)$ via $\begin{pmatrix} A & B \\ -B & A \end{pmatrix}$. The map

$$Mp(n,\mathbb{R})_c \longrightarrow \mathbb{H}_n, \qquad P \longmapsto q_c(P)\cdot(i\,I_n), \quad P \in MP(n,\mathbb{R})_c$$

gives the coset space of $Mp(n,\mathbb{R})_c$ with respect to $q_c^{-1}(U(n))$, i.e.,

$$Mp(n,\mathbb{R})_c / q_c^{-1}(U(n)) = \mathbb{H}_n.$$

For $\Omega = i\,I_n$, the map $B_{c;iI_n} : q_c^{-1}(U(n)) \longrightarrow \mathbb{C}_1^*$ defined by

$$B_{c;iI_n}(P) := B_c(P; i\,I_n), \quad P \in q_c^{-1}(U(n)) \tag{2.4.14}$$

is a continuous character. If $P \in q_c^{-1}(U(n))$ with $q_c(P) = \begin{pmatrix} A & B \\ -B & A \end{pmatrix} \in U(n)$, then

$$\left\{B_{c;iI_n}(P)\right\}^2 = \left\{B_c(P;iI_n)\right\}^2 = \left\{\det(A - i\,B)\right\}^{-m}.$$

We define the map $\det_c^* : U(n) \longrightarrow \mathbb{C}_1^*$ by

$$\det{}_c^*\begin{pmatrix} A & B \\ -B & A \end{pmatrix} := \{\det(A - i\,B)\}^{-m}, \quad \begin{pmatrix} A & B \\ -B & A \end{pmatrix} \in U(n). \tag{2.4.15}$$

and the map $Sq : \mathbb{C}_1^* \longrightarrow \mathbb{C}_1^*$ by $Sq(\alpha) = \alpha^2$ with $\alpha \in \mathbb{C}_1^*$. The we have the following commutative diagram :

$$\begin{array}{ccc} q_c^{-1}(U(n)) & \xrightarrow{B_{c;iI_n}} & \mathbb{C}_1^* \\ {\scriptstyle q_c}\big\downarrow & & \big\downarrow{\scriptstyle Sq} \\ U(n) & \xrightarrow{\det_c^*} & \mathbb{C}_1^* \end{array}$$

diagram 14.1

Thus $q_c^{-1}(U(n))$ along with its topology is the fibre product of $\det_c^*$ and Sq. Since $U(n)$ and $\mathbb{C}_1^*$ are connected, $q_c^{-1}(U(n))$ is connected. □

Corollary 3. *The exact sequence*

$$1 \longrightarrow \{\pm 1\} \longrightarrow Mp(n,\mathbb{R})_c \xrightarrow{q_c} Sp(n,\mathbb{R}) \longrightarrow 1 \tag{2.4.16}$$

is non-split and $[Mp(n,\mathbb{R})_c, Mp(n,\mathbb{R})_c] = Mp(n,\mathbb{R})_c$.

Proof. Embed $U(1)$ into $U(n)$ via $z \mapsto \mathrm{diag}(z,1,1,\cdots,1)$, and embedd $U(n)$ into $Sp(n,\mathbb{R})$ via

$$U(n) \ni A + iB \longmapsto \begin{pmatrix} A & B \\ -B & A \end{pmatrix} \in Sp(n,\mathbb{R}) \quad \text{with } A, B \in \mathbb{R}^{(n,n)}.$$

So $U(1) \subset U(n) \subset Sp(n,\mathbb{R})$. According to the commutative diagram in the proof of Proposition 10, the exact sequence

$$1 \longrightarrow \{\pm 1\} \longrightarrow q_c^{-1}(U(1)) \xrightarrow{q_c} U(1) \longrightarrow 1$$

can be identified to

$$1 \longrightarrow \{\pm 1\} \longrightarrow \mathbb{C}_1^* \xrightarrow{Sq} \mathbb{C}_1^* \longrightarrow 1. \tag{2.4.17}$$

If we restrict the exact sequence (2.4.17) to the torsion subgroups, then we get the non-split exact sequence

$$1 \longrightarrow \mathbb{Z}/2\mathbb{Z} \longrightarrow \mathbb{Q}/\mathbb{Z} \xrightarrow{m_2} \mathbb{Q}/\mathbb{Z} \longrightarrow 1, \tag{2.4.18}$$

where $m_2 : \mathbb{Q}/\mathbb{Z} \longrightarrow \mathbb{Q}/\mathbb{Z}$ is the map defined by $m_2(x) = 2\,x$ for $x \in \mathbb{Q}/\mathbb{Z}$. Thus the exact sequence (2.4.16) is non-split.

For brevity, we put $Mp_{(c)} := Mp(n,\mathbb{R})_c$. Since $[Sp(n,\mathbb{R}), Sp(n,\mathbb{R})] = Sp(n,\mathbb{R})$, $[Mp_{(c)}, Mp_{(c)}]$ sits in the exact sequence

$$1 \longrightarrow \{\pm 1\} \cap [Mp_{(c)}, Mp_{(c)}] \longrightarrow [Mp_{(c)}, Mp_{(c)}] \longrightarrow Sp(n,\mathbb{R}) \longrightarrow 1. \tag{2.4.19}$$

Assume $\{\pm 1\} \cap [Mp_{(c)}, Mp_{(c)}]$ is trivial. Then according to the above exact sequence (2.4.19), we have an isomorphism $\phi : Sp(n,\mathbb{R}) \longrightarrow [Mp_{(c)}, Mp_{(c)}] \neq Mp_{(c)}$. Thus the exact sequence (2.4.16) is split because $q_c \circ \phi$ the identity map. This contradicts the fact that the exact sequence (2.4.16) is non-split. Hence we obtain

$$\{\pm 1\} \cap [Mp_{(c)}, Mp_{(c)}] = \{\pm 1\} \qquad \text{and} \qquad [Mp_{(c)}, Mp_{(c)}] = Mp_{(c)}.$$

□

Corollary 4. *For a fixed element $\Omega \in \mathbb{H}_n$, we let*

$$U(\Omega) = \big\{\, M \in Sp(n,\mathbb{R}) \,\big|\, M \cdot \Omega = \Omega \,\big\}.$$

Let $M_\Omega \in Sp(n,\mathbb{R})$ such that $\Omega = M \cdot (iI_n)$. Then $U(\Omega) = M_\Omega\, U(n) M_\Omega^{-1}$. If $P \in q_c^{-1}(U(\Omega))$ such that $q_c(P) = M_\Omega\, q_c(P_0) M_\Omega^{-1}$ with $P_0 \in q_c^{-1}(U(n))$, then

$$\big\{B_c(P;\Omega)\big\}^2 = \det{}_c^*\big(q_c(P)\big),$$

where $\det_c^* : U(n) \longrightarrow \mathbb{C}_1^*$ is the map defined by Formula (2.4.15).

Proof. The case $\Omega = i\,I_n$ has already been proved before. We note that $U(iI_n) = U(n)$. For $M = \begin{pmatrix} A & B \\ C & D \end{pmatrix}$ and $\Omega \in \mathbb{H}_n$, we put $J(M,\Omega) = \det(C\Omega + D)$. By definition, if $P \in q_c^{-1}(U(\Omega))$ such that $q_c(P) = M_\Omega\, q_c(P_0) M_\Omega^{-1}$ with $P_0 \in q_c^{-1}(U(n))$ and $q_c(P_0) = \begin{pmatrix} A & B \\ -B & A \end{pmatrix} \in U(n)$, then

$$\begin{aligned}
\big\{B_c(P;\Omega)\big\}^2 &= J(q_c(P),\Omega)^{-m} = \big\{J\big(M_\Omega\, q_c(P_0) M_\Omega^{-1},\Omega\big)\big\}^{-m} \\
&= \big\{J(M_\Omega\, q_c(P_0), iI_n)\, J(M_\Omega^{-1},\Omega)\big\}^{-m} \\
&= \big\{J(M_\Omega, iI_n)\, J(M_\Omega^{-1},\Omega)\, J(q_c(P_0), iI_n)\big\}^{-m} \\
&= \big\{J(q_c(P_0), iI_n)\big\}^{-m} \\
&= \big\{\det(A - iB)\big\}^{-m} = \det{}_c^*\big(q_c(P_0)\big).
\end{aligned}$$

□

2.5 Theta Series with Quadratic Forms

In this chapter, we review the theta series of several type.

Definition 11. A symmetric integral matrix S of degree m is said to be even if ${}^t\xi\, S\, \xi \equiv 0 \bmod 2$ for all $\xi \in \mathbb{Z}^{(m,1)}$. The level q of an even symmetric nonsingular matrix S is defined to be the smallest positive integer such that qS^{-1} is even.

It is well known that if S is positive definite even integral matrix of degree m such that $\det S = 1$, then m is divisible by 8.

Definition 12. For a symmetric integral matrix T of degree n and a symmetric integral matrix S of degree m, we define

$$A(S,T) := \sharp\{\, \xi \in \mathbb{Z}^{(m,n)} \mid {}^t\xi\, S\, \xi \, = T\,\}.$$

We observe that if S is positive definite, $A(S,T)$ is finite. It is easy to see that S_1 and S_2 are equivalent, that is, ${}^tUS_1U = S_2$ for some $U \in GL(m,\mathbb{Z})$ if and only if $A(S_1,T) = A(S_2,T)$ for all n and symmetric integral matrices T of degree n.

Let S be a positive definite integral matrix of degree m. We define the theta series $\vartheta_S : \mathbb{H}_n \longrightarrow \mathbb{C}$ by

$$\vartheta_S(\Omega) = \sum_{\xi \in \mathbb{Z}^{(m,n)}} e^{\pi\, i\, \sigma(S\,\xi\,\Omega\,{}^t\xi)}, \qquad \Omega \in \mathbb{H}_n. \tag{2.5.1}$$

Then $\vartheta_S(\Omega)$ is a holomorphic function on $\mathbb{H}_n$. We see that

$$\vartheta_S(\Omega) = \sum_{T={}^tT \geq 0} A(S,T)\, e^{\pi\, i\, \sigma(T\Omega)},$$

where T runs over the set of all semipositive symmetric integral matrices of degree n.

Theorem 18. *Let S be a positive definite symmetric integral matrix of degree m. Then $\vartheta_S(\Omega)$ satisfies the transformation formula*

$$\vartheta_{S^{-1}}(-\Omega^{-1}) = (\det S)^{\frac{n}{2}} \left(\det \frac{\Omega}{i}\right)^{\frac{m}{2}} \vartheta_S(\Omega) \qquad \text{for all } \Omega \in \mathbb{H}_n. \tag{2.5.2}$$

Here the function $h : \mathbb{H}_n \longrightarrow \mathbb{C}$ given by

$$h(\Omega) = \left(\det \frac{\Omega}{i}\right)^{\frac{1}{2}}, \qquad \Omega \in \mathbb{H}_n$$

is the function determined uniquely by the following properties

(a) $h^2(\Omega) = \left(\det \frac{\Omega}{i}\right), \quad \Omega \in \mathbb{H}_n,$

(b) $h(iY) = (\det Y)^{\frac{1}{2}}$ for any positive definite symmetric real matrix Y of degree n.

For a positive integer m, we define

$$\left(\det \frac{\Omega}{i}\right)^{\frac{m}{2}} = \left\{\left(\det \frac{\Omega}{i}\right)^{\frac{1}{2}}\right\}^m, \qquad \Omega \in \mathbb{H}_n.$$

Proof. For a fixed element $\Omega \in \mathbb{H}_n$, we define $f : \mathbb{R}^{(m,n)} \longrightarrow \mathbb{C}$ by

$$f(x) = \sum_{\xi \in \mathbb{Z}^{(m,n)}} e^{\pi\, i\, \sigma(S(\xi+x)\,\Omega\, {}^t(\xi+x))}, \quad x \in \mathbb{R}^{(m,n)}. \tag{2.5.3}$$

We observe that f is well defined because the sum of the right hand side of (2.5.3) converges absolutely. It is clear that if $x = (x_{ij})$ is a coordinate in $\mathbb{R}^{(m,n)}$, then f is periodic in x_{ij} with period 1. That is,

$$f(x+\alpha) = f(x) \qquad \text{for all } \alpha \in \mathbb{Z}^{(m,n)}.$$

Thus f has the Fourier series

$$f(x) = \sum_{\alpha \in \mathbb{Z}^{(m,n)}} c_\alpha\, e^{2\,\pi\, i\, \sigma(x\, {}^t\alpha)}, \quad x \in \mathbb{R}^{(m,n)}, \tag{2.5.4}$$

where

$$\begin{aligned} c_\alpha &= \int_0^1 \cdots \int_0^1 f(y)\, e^{-2\,\pi\, i\, \sigma(y\, {}^t\alpha)}\, dy \\ &= \int_0^1 \cdots \int_0^1 \sum_{\xi \in \mathbb{Z}^{(m,n)}} e^{\pi\, i\, \sigma\{S(\xi+y)\,\Omega\, {}^t(\xi+y)\}} \cdot e^{-2\,\pi\, i\, \sigma(y\, {}^t\alpha)}\, dy \\ &= \int_{\mathbb{R}^{(m,n)}} e^{\pi\, i\, \sigma(S\, y\, \Omega\, {}^t y - 2\, y\, {}^t\alpha)}\, dy \\ &= \int_{\mathbb{R}^{(m,n)}} e^{\pi\, i\, \sigma\{S(y\,\Omega\, {}^t y - 2\, y\, {}^t(S^{-1}\,\alpha))\}}\, dy \\ &= (\det S)^{-\frac{n}{2}} \left(\det \frac{\Omega}{i}\right)^{-\frac{m}{2}} e^{-\pi\, i\, \sigma(S^{-1}\,\alpha\,\Omega^{-1}\, {}^t\alpha)} \quad \text{(by Lemma 34)}. \end{aligned}$$

According to Formulas (2.5.3) and (2.5.4),

$$\begin{aligned}\vartheta_S(\Omega) = f(0) &= \sum_{\alpha\in\mathbb{Z}^{(m,n)}} c_\alpha \\ &= (\det S)^{-\frac{n}{2}} \left(\det \frac{\Omega}{i}\right)^{-\frac{m}{2}} e^{\pi\, i\,\sigma\{S^{-1}\alpha\,(-\Omega^{-1})\,{}^t\alpha\}} \\ &= (\det S)^{-\frac{n}{2}} \left(\det \frac{\Omega}{i}\right)^{-\frac{m}{2}} \vartheta_{S^{-1}}(-\Omega^{-1}).\end{aligned}$$

Consequently we obtain the formula (2.5.2). □

Let S be an positive definite even integral symmetric matrix of degree m. Let A and B be $m\times n$ rational matrices. We define the theta series

$$\vartheta_{S;A,B}(\Omega) = \sum_{\xi\in\mathbb{Z}^{(m,n)}} e^{\pi\, i\,\sigma\{S(\xi+\frac{1}{2}A)\,\Omega\,{}^t(\xi+\frac{1}{2}A)\,+\,{}^tB\,\xi\}}.$$

Theorem 19. *Let S be an positive definite even integral symmetric matrix of degree m. Let A and B be $m\times n$ rational matrices. Then $\vartheta_{S;A,B}(\Omega)$ satisfies the transformation formula*

$$\vartheta_{S^{-1};A,B}(-\Omega^{-1}) = e^{-\frac{1}{2}\pi\, i\,\sigma({}^tAB)}\,(\det S)^{\frac{n}{2}} \left(\det \frac{\Omega}{i}\right)^{\frac{m}{2}} \vartheta_{S;B,-A}(\Omega) \quad (2.5.5)$$

for all $\Omega\in\mathbb{H}_n$.

Proof. Following the argument of the proof of Theorem 18, we can obtain the formula (2.5.5). We leave the detail to the reader. □

Definition 13. A holomorphic function $f:\mathbb{H}_n\longrightarrow\mathbb{C}$ is called a Siegel modular form of weight $k\in\mathbb{Z}$ if it satisfies the following properties:

1) $f(M\cdot\Omega) = \det(C\Omega+D)^k f(\Omega)$ for $\gamma = \begin{pmatrix} A & B \\ C & D\end{pmatrix}\in\Gamma_n$.

2) f is bounded in the domain $Y\geq Y_0>0$ with $\Omega = X+i\,Y,\ X,Y$ real.

We will give some examples of Siegel modular forms using the so-called thetanullwerte. For $a,b\in\mathbb{Z}^n$, we consider the thetanullwerte

$$\vartheta(\Omega\,;a,b) = \sum_{\xi\in\mathbb{Z}^n} e^{\pi\, i\,\sigma\{\,{}^t(\xi+\frac{1}{2}a)\,\Omega\,(\xi+\frac{1}{2}a)\,+\,{}^tb\,\xi\,\}},\quad \Omega\in\mathbb{H}_n. \quad (2.5.6)$$

Lemma 37. *Let $a,b\in\mathbb{Z}^n$. Then $\vartheta(\Omega\,;a,b)$ satisfies the following properties*

(a) $\vartheta(\Omega\,;a,b_1) = \vartheta(\Omega\,;a,b_2)$ *if* $b_1\equiv b_2$ *mod 2.*

(b) If $\tilde{a}\in\mathbb{Z}^n$*, then* $\vartheta(\Omega\,;a+2\,\tilde{a},b) = (-1)^{{}^tb\,\tilde{a}}\,\vartheta(\Omega\,;a,b)$.

(c) $\vartheta(\Omega\,;a,b) = (-1)^{{}^tab}\,\vartheta(\Omega\,;a,b)$.

(d) $\vartheta(\Omega\,;a,b) = 0$ *if* ${}^tab\not\equiv 0$ *mod 2.*

Proof. (a) follows from a direct computation. If we put $\xi_* = \xi + \widetilde{a}$,

$$\begin{aligned}\vartheta(\Omega\,;a+2\,\widetilde{a},b) &= \sum_{\xi\in\mathbb{Z}^n} e^{\pi\, i\,\sigma\{\,{}^t(\xi+\frac{1}{2}a+\widetilde{a})\,\Omega\,(\xi+\frac{1}{2}a+\widetilde{a})\,+\,{}^tb\,\xi\,\}} \\ &= \sum_{\xi_*\in\mathbb{Z}^n} e^{\pi\, i\,\sigma\{\,{}^t(\xi_*+\frac{1}{2}a)\,\Omega\,(\xi_*+\frac{1}{2}a)\,+\,{}^tb(\xi_*-\widetilde{a})\,\}} \\ &= e^{-\pi\, i\,{}^tb\widetilde{a}}\,\vartheta(\Omega\,;a,b).\end{aligned}$$

Therefore we get the formula (b). If we substitute ξ into $-\xi - a$, we obtain the formula (c). (d) follows immediately from the formula (c). □

A pair $\{a, b\}$ with $a, b \in \{0,1\}^n$ is called a **theta characteristic**. A theta characteristic $\{a, b\}$ is said to be **even** (resp. **odd**) if ${}^tab$ is even (resp. odd). By induction on n, we can show that the number of even theta characteristics is $(2^n+1)\,2^{n-1}$.

Let $\gamma = \begin{pmatrix} A & B \\ C & D \end{pmatrix} \in \Gamma_n$ and let $\{a, b\}$ be a theta characteristic. We define

$$\gamma \diamond \begin{pmatrix} a \\ b \end{pmatrix} :\equiv \begin{pmatrix} D & -C \\ -B & A \end{pmatrix}\begin{pmatrix} a \\ b \end{pmatrix} + \begin{pmatrix} (C\,{}^tD)_0 \\ (A\,{}^tB)_0 \end{pmatrix} \mod 2, \tag{2.5.7}$$

where T_0 is the column vector determined by the diagonal entries of an $n \times n$ matrix T.

Theorem 20. *(1) The Siegel modular group Γ_n acts on the set $\mathscr{C}$ of theta characteristics by*

$$\begin{pmatrix} a \\ b \end{pmatrix} \mapsto \gamma \diamond \begin{pmatrix} a \\ b \end{pmatrix}, \quad \gamma \in \Gamma_n,\ \{a,b\} \in \mathscr{C}.$$

(2) The sign $(-1)^{{}^tab}$ of the theta characteristic $\{a, b\}$ is invariant under the action (2.5.7) of Γ_n.

(3) Γ_n acts on the set $\mathscr{C}^e$ of all even theta characteristics transitively.

(4) If $\gamma = \begin{pmatrix} A & B \\ C & D \end{pmatrix} \in \Gamma_n$, $\Omega \in \mathbb{H}_n$ and $\{a,b\} \in \mathscr{C}$, then we have

$$\vartheta^2(\gamma\cdot\Omega\,;a,b) = \nu(\gamma)\,\det(C\Omega + D)\,\vartheta^2\big(\Omega\,;\widetilde{a},\widetilde{b}\big), \tag{2.5.8}$$

where

(a) $\nu(\gamma)^4 = 1$,

(b) $\begin{pmatrix} \widetilde{a} \\ \widetilde{b} \end{pmatrix} = \gamma \diamond \begin{pmatrix} a \\ b \end{pmatrix}$.

Proof. By a direct computation, we prove the statement (a). It suffices to show the invariance of the sign of $(-1)^{{}^tab}$ under the generators $t_S = \begin{pmatrix} I_n & S \\ 0 & I_n \end{pmatrix}$ with

$S = {}^tS \in \mathbb{Z}^{(n,n)}$ and J_n. By a simple computation,

$$\begin{pmatrix} I_n & S \\ 0 & I_n \end{pmatrix} \diamond \begin{pmatrix} a \\ b \end{pmatrix} \equiv \begin{pmatrix} a \\ b - S\,a + S_0 \end{pmatrix} \mod 2,$$
$$\begin{pmatrix} I_n & S \\ 0 & I_n \end{pmatrix} \diamond \begin{pmatrix} a \\ b \end{pmatrix} \equiv \begin{pmatrix} b \\ -a \end{pmatrix} \mod 2.$$

Therefore it is obvious that the sign of $(-1)^{{}^tab}$ is invariant under the actions of t_S and J_n. In order to prove the transitivity of Γ_n on $\mathscr{C}^e$, first of all we have to prove the fact that given an even characteristic $\{a,b\} \in \mathscr{C}^e$, there exists an element $\gamma = \begin{pmatrix} A & B \\ C & D \end{pmatrix} \in \Gamma_n$ such that

$$\gamma \diamond \begin{pmatrix} 0 \\ 0 \end{pmatrix} = \begin{pmatrix} a \\ b \end{pmatrix}, \quad i.e., \quad a \equiv (C\,{}^tD)_0, \; b \equiv (A\,{}^tB)_0 \mod 2.$$

We decompose

$$a = \begin{pmatrix} a_1 \\ a_2 \end{pmatrix}, \quad a_1 \in \mathbb{Z}, \; a_2 \in \mathbb{Z}^{n-1},$$
$$b = \begin{pmatrix} b_1 \\ b_2 \end{pmatrix}, \quad b_1 \in \mathbb{Z}, \; b_2 \in \mathbb{Z}^{n-1}.$$

Case 1. $a_1\, b_1 = 0$

Then $\{a_1, b_1\}$ is even and also $\{a_2, b_2\}$ is even. By induction on n, we can find $\gamma \in \Gamma_n$ such that $\gamma \diamond \begin{pmatrix} 0 \\ 0 \end{pmatrix} = \begin{pmatrix} a \\ b \end{pmatrix}$.

Case 2. $a_1 = b_1 = 1$

Since ${}^tab$ is even, there exists an index ν with $2 \leq \nu \leq n$ such that $a_\nu = b_\nu = 1$. Therefore we can find a symmetric integral matrix $S = tS \in \mathbb{Z}^{(n,n)}$ so that

$$\begin{pmatrix} I_n & S \\ 0 & I_n \end{pmatrix} \diamond \begin{pmatrix} a \\ b \end{pmatrix} \equiv \begin{pmatrix} a \\ b - S\,a + S_0 \end{pmatrix} \mod 2$$

is an even theta characteristic satisfying the assumption of Case 1.

According to Case 1 and Case 2, we see that Γ_n acts on $\mathscr{C}^e$ transitively.

The transformation formula (2.5.8) for the generator J_n follows from the formula (2.5.5) with $S = 1$ and $m = 1$. For a generator t_S with $S = {}^tS \in \mathbb{Z}^{(n,n)}$, it is easy to see that

$$\vartheta(\Omega + S\,; a, b) = e^{\frac{\pi i}{4}\,{}^taSa}\, \vartheta(\Omega\,; a, b + Sa + S_0). \tag{2.5.9}$$

In fact, (2.5.9) follows from the following simple fact that ${}^t\xi S\xi \equiv {}^tS_0\,\xi \bmod 2$ for any $\xi \in \mathbb{Z}^n$ and $x^2 \equiv x \bmod 2$ for any $x \in \mathbb{Z}$. We know that $\vartheta(\Omega\,;0,0) \neq 0$ because $\vartheta(i\,Y\,;0,0) > 0$. □

Theorem 21. *We set*

$$k_n = \begin{cases} 8 & \text{if} \quad n = 1 \\ 2 & \text{if} \quad n = 2 \\ 1 & \text{if} \quad n \geq 3. \end{cases}$$

We define the function $\Delta^{(n)}(\Omega)$ on $\mathbb{H}_n$ by

$$\Delta^{(n)}(\Omega) := \prod_{\{a,b\}} \vartheta(\Omega\,;a,b)^{k_n},$$

where $\{a,b\}$ runs through even theta characteristics. Then $\Delta^{(n)}(\Omega)$ is a nonzero Siegel modular form on $\mathbb{H}_n$ of weight 12, 10 and $(2^n+1)\,2^{n-2}$ respectively if $n=1,2$ and $n \geq 3$ respectively.

Proof. The proof can be found in [7]. □

Theorem 22. *Let m be an even positive integer. Let S be a positive definite even integral symmetric matrix of degree m and of level q. Then for all $\gamma = \begin{pmatrix} A & B \\ C & D \end{pmatrix} \in \Gamma_{n,0}(q)$ with $\det D > 0$,*

$$\vartheta_S(\gamma \cdot \Omega) = \nu_S(\gamma)\, \det(C\Omega + D)^{\frac{m}{2}}\, \vartheta_S(\Omega), \qquad \Omega \in \mathbb{H}_n,$$

where

$$\begin{aligned} \nu_S(\gamma) &= (\det D)^{\frac{m}{2}-mn} \sum_{\xi \in \mathbb{Z}^{(m,n)}} e^{\pi\, i\, \sigma(BD^{-1}\,{}^t\xi\, S\, \xi)} \\ &= (sgn\, \det D)^{\frac{m}{2}} \left(\frac{(-1)^{\frac{m}{2}}\, \det S}{|\det D|} \right). \end{aligned}$$

Here $\left(\frac{a}{b}\right)$ denotes the generalized Legendre symbol.

Proof. The proof can be found in [7], pp. 302-303. □

Theorem 23. *Let m be an even positive integer. Let S be a positive definite even integral symmetric matrix of degree m and of level q. Then $\vartheta_S(\Omega)$ is a modular form with respect to the principal congruence subgroup $\Gamma_n(q)$ of Γ_n.*

Proof. The proof follows from Theorem 22. □

Theorem 24. *Let S be an positive definite even integral symmetric matrix of degree m. Let A and B be $m \times n$ rational matrices. Then the theta series*

$$\vartheta_{S;A,B}(\Omega) = \sum_{\xi \in \mathbb{Z}^{(m,n)}} e^{\pi\, i\, \sigma\{S(\xi + \frac{1}{2}A)\,\Omega\,{}^t(\xi + \frac{1}{2}A) + {}^tB\,\xi\}}$$

is a modular form of weight $\frac{m}{2}$ *with respect to a certain congruence subgroup* $\Gamma_n(\ell)$ *of* Γ_n.

Proof. The proof can be found in [1]. □

2.6 Theta Series in Spherical Harmonics

Let S be a positive definite symmetric $m \times m$ rational matrix, and let α and β be an $m \times n$ rational matrix. We define the theta series

$$\vartheta_S \begin{bmatrix} \alpha \\ \beta \end{bmatrix} : \mathbb{H}_n \times \mathbb{C}^{(m,n)} \longrightarrow \mathbb{C}$$

by

$$\vartheta_S \begin{bmatrix} \alpha \\ \beta \end{bmatrix} (\Omega, Z) := \sum_{N \in \mathbb{Q}^{(m,n)}} \chi \begin{bmatrix} \alpha \\ \beta \end{bmatrix} (N)\, e^{\pi\, i\, \sigma({}^tNSN\Omega + 2\, {}^tNZ)}, \tag{2.6.1}$$

where $\Omega \in \mathbb{H}_n$, $Z \in \mathbb{C}^{(m,n)}$ and

$$\chi \begin{bmatrix} \alpha \\ \beta \end{bmatrix} (N) = \begin{cases} 1 & \text{if } N - \alpha \notin \mathbb{Z}^{(m,n)} \\ e^{2\,\pi\, i\, \sigma({}^tNB)} & \text{otherwise.} \end{cases}$$

Let $\mathfrak{P}_{m,n}$ be the algebra of complex valued polynomial functions on $\mathbb{C}^{(m,n)}$. We take a coordinate $Z = (z_{kj})$ in $\mathbb{C}^{(m,n)}$.

Definition 14. Let S, α and β be as above. For a homogeneous polynomial $P \in \mathfrak{P}_{m,n}$, we define

$$\vartheta_{S,P} \begin{bmatrix} \alpha \\ \beta \end{bmatrix} (\Omega, Z) = \sum_{N \in \mathbb{Q}^{(m,n)}} \chi \begin{bmatrix} \alpha \\ \beta \end{bmatrix} (N)\, P(N)\, e^{\pi\, i\, \sigma({}^tNSN\Omega + 2\, {}^tNZ)}, \tag{2.6.2}$$

$$\vartheta_{S,P}(\Omega, Z) = \vartheta_{S,P} \begin{bmatrix} 0 \\ 0 \end{bmatrix} (\Omega, Z), \tag{2.6.3}$$

$$\vartheta_{S,P} \begin{bmatrix} \alpha \\ \beta \end{bmatrix} (\Omega) = \vartheta_{S,P} \begin{bmatrix} \alpha \\ \beta \end{bmatrix} (\Omega, 0). \tag{2.6.4}$$

For any homogeneous polynomial P in $\mathfrak{P}_{m,n}$, we put

$$P(\partial) = P\left(\frac{\partial}{\partial z_{kj}} \right), \quad 1 \le k \le m,\ 1 \le j \le n.$$

Then we get

$$\begin{aligned} & P(\partial) \vartheta_S \begin{bmatrix} \alpha \\ \beta \end{bmatrix} (\Omega, Z) \qquad (2.6.5) \\ = & \sum_{N \in \mathbb{Q}^{(m,n)}} \chi \begin{bmatrix} \alpha \\ \beta \end{bmatrix} (N)\, P(2\,\pi i N)\, e^{\pi\, i\, \sigma({}^tNSN\Omega + 2\, {}^tNZ)}. \end{aligned}$$

Definition 15. Let $T = (t_{kj})$ be the inverse matrix of S. Then a polynomial P in $\mathfrak{P}_{m,n}$ is said to be pluriharmonic with respect to S if it satisfies the equations

$$\sum_{k,l=1}^{m} \frac{\partial^2 P}{\partial z_{ki}\partial z_{lj}}\, t_{kl} = 0 \qquad \text{for all } i,j = 1,2,\cdots,n.$$

Theorem 25. *Let S, α and β be as above. Then for all $\begin{pmatrix} A & B \\ C & D \end{pmatrix}$ in a suitable subgroup Γ of Γ_n, we have*

$$\vartheta_S \begin{bmatrix} \alpha \\ \beta \end{bmatrix} \left((A\Omega + B)(C\Omega + D)^{-1}, Z(C\Omega + D)^{-1} \right) \tag{2.6.6}$$
$$= \det(C\Omega + D)^{\frac{m}{2}}\, e^{\pi\, i\, \sigma(Z(C\Omega+D)^{-1}C\,{}^tZS^{-1})}\, \vartheta_S \begin{bmatrix} \alpha \\ \beta \end{bmatrix} (\Omega, Z)$$

Proof. $Sp(n,\mathbb{R})$ acts on the homogeneous space $\mathbb{H}_n \times \mathbb{C}^{(m,n)}$ by

$$M \cdot (\Omega, Z) = \left((A\Omega + B)(C\Omega + D)^{-1},\ Z(C\Omega + D)^{-1} \right),$$

where $M = \begin{pmatrix} A & B \\ C & D \end{pmatrix} \in Sp(n,\mathbb{R})$, $\Omega \in \mathbb{H}_n$ and $Z \in \mathbb{C}^{(m,n)}$. It is known that $Sp(n,\mathbb{R})$ is generated by the translations t_b with $b = {}^tb$ and the inversion σ_n. Thus it suffices to prove the functional equation (2.6.6) for the generators t_b and σ_n in a suitable congruence subgroup Γ of Γ_n.

For $t_b = \begin{pmatrix} I_n & b \\ 0 & I_n \end{pmatrix} \in Sp(n,\mathbb{R})$,

$$\vartheta_S \begin{bmatrix} \alpha \\ \beta \end{bmatrix} (\Omega + b, Z) = \sum_{N \in \mathbb{Q}^{(m,n)}} \chi \begin{bmatrix} \alpha \\ \beta \end{bmatrix} (N)\, e^{\pi\, i\, \sigma({}^tNSN(\Omega+b)\,+\,2\,{}^tNZ)}$$
$$= \vartheta_S \begin{bmatrix} \alpha \\ \beta \end{bmatrix} (\Omega, Z)$$

if we choose suitable b's so that $e^{\pi\, i\, \sigma({}^tNSNb)} = 1$. This is possible because S, α and β are *rational* matrices.

For the inversion $\sigma_n = \begin{pmatrix} 0 & -I_n \\ I_n & 0 \end{pmatrix}$, we can prove the functional equation (2.6.6) following the argument in the proof of Theorem 18. We leave the details to a reader. Another representation theoretic proof can be found in [28] □

Lemma 38. *Let $\mathfrak{P}_N := \mathbb{C}[X_1, \cdots, X_N]$. For $P \in \mathfrak{P}_N$, we let $P(\partial)$ denote the differential operator $P\left(\frac{\partial}{\partial X_1}, \cdots, \frac{\partial}{\partial X_N}\right)$. For P, $Q \in \mathfrak{P}_N$, we define*

$$\langle P, Q\rangle = (P(\partial)Q)(0).$$

Then $\langle\ ,\ \rangle$ is a symmetric nondegenerate bilinear form on $\mathfrak{P}_N$ which satisfies the property $\langle P, QR\rangle = \langle Q(\partial)P, R\rangle = \langle R(\partial)P, Q\rangle$ for all $P, Q, R \in \mathfrak{P}_N$.

Proof. We first observe that

$$\left\langle X_1^{a_1}\cdots X_N^{a_N}, X_1^{b_1}\cdots X_N^{b_N}\right\rangle = \begin{cases} a_1!\cdots a_N! & \text{if } (a_1,\cdots,a_N) = (b_1,\cdots,b_N), \\ 0 & \text{otherwise.}\end{cases}$$

Thus $\langle P, Q\rangle$ is a symmetric nondegenerate bilinear form on $\mathfrak{P}_N$. Similarly $\langle P, QR\rangle = \langle Q(\partial)P, R\rangle = \langle R(\partial)P, Q\rangle$ is easily shown for monomials P, Q, R. Hence we complete the proof. □

Lemma 39. *Let $\mathfrak{H}(S) \subset \mathfrak{P}_{m,n}$ be the space of pluriharmonic polynomials with respect to S, and $I \subset \mathfrak{P}_{m,n}$ be the ideal generated by the $h_{ij} = \sum_{k,l=1}^m t_{kl}z_{ki}z_{lj}$ for all i, $j = 1,\cdots,n$, where $T = (t_{kl}) = S^{-1}$ as before in Definition 16.2. Then $\mathfrak{H}(S) = I^\perp$ with respect to the pairing $\langle\ ,\ \rangle$ introduced in Lemma 16.4, and*

$$\mathfrak{P}_{m,n} = \mathfrak{H}(S) \oplus I \qquad (\textit{orthogonal sum}).$$

Proof. Let $P \in \mathfrak{P}_{m,n}$. Then $\langle fh_{ij}, P\rangle = \big(f(\partial)h_{ij}(\partial)P\big)(0) = 0$ for all $f \in \mathfrak{P}_{m,n}$ if and only if $h_{ij}(\partial)P = 0$ for all i, j if and only if P is pluriharmonic with respect to S. Thus $\mathfrak{H}(S) = I^\perp$. Let $\mathfrak{P}_{m,n}(\mathbb{R}) = \mathbb{R}[Z_{11}, Z_{12},\cdots,Z_{mn}]$. By the same argument, we have $\mathfrak{H}(S)_\mathbb{R} = I_\mathbb{R}^\perp$, where $\mathfrak{H}(S)_\mathbb{R} = \mathfrak{H}(S)\cap\mathfrak{P}_{m,n}(\mathbb{R})$ and $I_\mathbb{R} = I\cap\mathfrak{P}_{m,n}(\mathbb{R})$. It is easy to see that $\langle\ ,\ \rangle$ is positive definite on $\mathfrak{P}_{m,n}(\mathbb{R})$. So $\mathfrak{P}_{m,n}(\mathbb{R}) = \mathfrak{H}(S)_\mathbb{R}\oplus I_\mathbb{R}$. Therefore we have $\mathfrak{P}_{m,n} = \mathfrak{H}(S)\oplus I$. □

Lemma 40. *If P is a pluriharmonic polynomial in $\mathfrak{H}(S) \subset \mathfrak{P}_{m,n}$, then*

$$\left(P(\partial)\left[g(Z)\, e^{\sigma(ZC\,{}^tZS^{-1})}\right]\right)(0) = \big(P(\partial)g(Z)\big)(0) \tag{2.6.7}$$

for any $C \in \mathbb{C}^{(n,n)}$ and any analytic function g defined in a neighborhood of 0.

Proof. We put $h(Z) = \sigma(ZC\,{}^tZS^{-1})$ and $T = (t_{kl}) = S^{-1}$. It suffices to prove the formula (2.6.7) for any polynomials $g(Z)$.

$$\begin{aligned}\left(P(\partial)\left[g(Z)\,e^{h(Z)}\right]\right)(0) &= \sum_{n=0}^{\infty}\frac{1}{n!}\left(P(\partial)\left[g(Z)h(Z)^n\right]\right)(0)\\ &= \sum_{n=0}^{\infty}\frac{1}{n!}\langle P, gh^n\rangle\\ &= \sum_{n=0}^{\infty}\frac{1}{n!}\langle h(\partial)^n P, g\rangle \qquad \text{(by Lemma 38).}\end{aligned}$$

By the way, $h(\partial)P = 0$ because P is pluriharmonic. Indeed, if we put $C = (c_{ij})$ and $Z = (z_{ki})$, then we have

$$\begin{aligned}h(Z) &= \sigma(ZC\,{}^tZS^{-1})\\ &= \sum_{i,j=1}^{n} c_{ij}\left(\sum_{k,l=1}^{m} t_{kl}\,z_{ki}\,z_{lj}\right).\end{aligned}$$

We put $f_{ij}(Z) = \sum_{k,l=1}^{m} t_{kl}\,z_{ki}\,z_{lj}$. Then $h(\partial)P = \sum_{i,j=1}^{n} c_{ij}\left(f_{ij}(\partial)P\right) = 0$ because P is pluriharmonic. Therefore we get

$$\left(P(\partial)\left[g(Z)\,e^{h(Z)}\right]\right)(0) = \langle P, g\rangle = \left(P(\partial)g(Z)\right)(0).$$

□

Corollary 5. *If P is a pluriharmonic polynomial in $\mathfrak{H}(S) \subset \mathfrak{P}_{m,n}$ and C is an $n \times n$ symmetric complex matrix, then*

$$P(\partial)e^{\sigma(ZC\,{}^tZS^{-1})} = P(2\,C\,{}^tZS^{-1})\,e^{\sigma(ZC\,{}^tZS^{-1})}.$$

Proof. We put $h(Z) = e^{\sigma(ZC\,{}^tZS^{-1})}$. For any $A \in \mathbb{C}^{(m,n)}$, we let

$$f(Z) = h(Z+A) = h(Z)\,h(A)\,g(Z),$$

where $g(Z) = e^{2\,\sigma(AC\,{}^tZS^{-1})}$. Then

$$\begin{aligned}\left(P(\partial)h(Z)\right)(A) &= \left(P(\partial)f(Z)\right)(0)\\ &= h(A)\left(P(\partial)[h(Z)\,g(Z)]\right)(0)\\ &= h(A)\left(P(\partial)g(Z)\right)(0) \qquad \text{(Lemma 40).}\end{aligned}$$

But

$$\frac{\partial g}{\partial z_{ki}} = (2\,C\,{}^tA\,S^{-1})_{ki}\,g(Z).$$

By a repeated application of this, we have

$$P(\partial)g(Z) = P(2\,C\,{}^tA\,S^{-1})\,g(Z).$$

Therefore

$$\begin{aligned}\big(P(\partial)h(Z)\big)(A) &= h(A)\,\big(P(\partial)g(Z)\big)(0)\\ &= h(A)\,P(2\,C\,{}^t\!A\,S^{-1})\,g(0)\\ &= h(A)\,P(2\,C\,{}^t\!A\,S^{-1}).\end{aligned}$$

Hence $P(\partial)h(Z) = P(2\,C\,{}^t\!A\,S^{-1})\,h(Z)$. □

Lemma 41. *Let f be an analytic function on $\mathbb{C}^{(m,n)}$ and let $P \in \mathfrak{P}_{m,n}$. For $A \in \mathbb{C}^{(n,n)}$ and $B \in \mathbb{C}^{(m,m)}$, we let*

$$f_{A,B}(Z) = f(BZA) \qquad \text{and} \qquad P_{A,B}(Z) = P({}^tBZA^{-1}).$$

Then

$$P(\partial)f_{A,B}(Z) = \big(P_{A,B}(\partial)f\big)(BZA).$$

In particular, $\langle P, f_{A,B}\rangle = \langle P_{A,B}, f\rangle$.

Proof. We let $A = (a_{ij}) \in \mathbb{C}^{(n,n)}$, $b = (b_{kl}) \in \mathbb{C}^{(m,m)}$ and $Z = (z_{lp})$. By an easy computation, we get

$$\frac{\partial f_{A,B}}{\partial z_{lp}}(Z) = \sum_{k=1}^{m}\sum_{i=1}^{n} b_{kl}\,a_{pi}\,\frac{\partial f}{\partial z_{ki}}(BZA), \quad 1 \le l \le m,\ 1 \le p \le n.$$

We put

$$\widetilde{Z} = {}^tBZA^{-1} \quad \text{with } \widetilde{Z} = (\widetilde{z}_{lp}).$$

Since

$$\frac{\partial}{\partial \widetilde{z}_{lp}} = \sum_{k=1}^{m}\sum_{i=1}^{n} b_{kl}\,a_{pi}\,\frac{\partial}{\partial z_{ki}}, \quad 1 \le l \le m,\ 1 \le p \le n,$$

we have

$$\frac{\partial f_{A,B}}{\partial z_{lp}}(Z) = \frac{\partial f}{\partial \widetilde{z}_{lp}}(BZA) \qquad \text{for all } l, p.$$

Therefore we have $P(\partial)f_{A,B}(Z) = \big(P_{A,B}(\partial)f\big)(BZA)$. □.

Lemma 42. *$GL(n,\mathbb{C}) \times O(S)$ acts on $\mathfrak{H}_{m,n}$ by*

$$(A,B)P(Z) = P(B^{-1}ZA), \tag{2.6.8}$$

where $A \in GL(n,\mathbb{C})$, $B \in O(S)$ and $P \in \mathfrak{P}_{m,n}$. The space $\mathfrak{H}(S)$ of pluriharmonic polynomials in $\mathfrak{P}_{m,n}$ is invariant under the action (2.6.8).

Proof. According to Lemma 39, $I = \mathfrak{H}(S)^{\perp}$ is the ideal of $\mathfrak{P}_{m,n}$ generated by $h_{ij}(Z) = \sum_{k,l=1}^{m} t_{kl}\,z_{ki}\,z_{lj}$ for all i, j. So by Lemma 41, it suffices to show that $h_{ij}(ZA)$ and $h_{ij}(BZ)$ belong to I for all $A \in GL(n,\mathbb{C})$ and $B \in O(S)$. If $A = (a_{ij}) \in GL(n,\mathbb{C})$, $Z = (z_{ki}) \in \mathbb{C}^{(m,n)}$ and $T = (t_{kl}) = S^{-1}$, then

$$\begin{aligned} h_{ij}(ZA) &= \sum_{p,q=1}^{n} a_{pi}\, a_{qj} \left(\sum_{k,l=1}^{m} t_{kl} z_{kp} z_{lq} \right) \\ &= \sum_{p,q=1}^{n} a_{pi}\, a_{qj}\, h_{pq}(Z) \in I. \end{aligned}$$

If $B = (b_{kl}) \in O(S)$, then

$$\begin{aligned} h_{ij}(BZ) &= \sum_{p,q=1}^{n} z_{pi}\, z_{qj} \left(\sum_{k,l=1}^{m} t_{kl} b_{kp} b_{lq} \right) \\ &= \sum_{p,q=1}^{n} z_{pi}\, z_{qj} \left({}^tBTB \right)_{pq}. \end{aligned}$$

Since $B \in O(S)$, we have $T = {}^tBTB$. Indeed $BS\,{}^tB =$ and hence ${}^tB\,S^{-1}\,B^{-1} = S^{-1}$. Thus ${}^tBTB = T$. Hence we have

$$h_{ij}(BZ) = \sum_{p,q=1}^{n} t_{pq}\, z_{pi}\, z_{qj} = h_{ij}(Z) \in I.$$

Therefore we complete the proof. □

Theorem 26. *Let S, α and β be as above. Let P be a pluriharmonic polynomial in $\mathfrak{P}_{m,n}$ with respect to S. Then*

$$\vartheta_{S,P}\begin{bmatrix}\alpha\\ \beta\end{bmatrix}(\Omega) = \det(C\Omega+D)^{-\frac{m}{2}}\, \vartheta_{S,\widetilde{P}}\begin{bmatrix}\alpha\\ \beta\end{bmatrix}\left((A\Omega+B)(C\Omega+D)^{-1}\right), \quad (2.6.9)$$

where $\widetilde{P}(Z) = P(Z(C\Omega+D))$, for all $\begin{pmatrix} A & B \\ C & D \end{pmatrix}$ in a suitable subgroup Γ of Γ_n.

Proof. Let P be a homogeneous pluriharmonic polynomial of degree k. Then according to Formula (2.6.5), we get

$$\begin{aligned} &(2\pi i)^{-k} P(\partial)\vartheta_S\begin{bmatrix}\alpha\\ \beta\end{bmatrix}(\Omega, Z) \\ &= (2\pi i)^{-k} \sum_{N\in\mathbb{Q}^{(m,n)}} \chi\begin{bmatrix}\alpha\\ \beta\end{bmatrix}(N)\, P(2\pi i N)\, e^{\pi\, i\,\sigma({}^tNSN\Omega + 2\,{}^tNZ)} \\ &= \sum_{N\in\mathbb{Q}^{(m,n)}} \chi\begin{bmatrix}\alpha\\ \beta\end{bmatrix}(N)\, P(N)\, e^{\pi\, i\,\sigma({}^tNSN\Omega + 2\,{}^tNZ)} \\ &= \vartheta_{S,P}\begin{bmatrix}\alpha\\ \beta\end{bmatrix}(\Omega, Z). \end{aligned}$$

Here the fact that P is homogeneous of degree k was used in the second equality. Putting $Z = 0$, we get

$$(2\pi i)^{-k} \left(P(\partial)\vartheta_S \begin{bmatrix} \alpha \\ \beta \end{bmatrix} \right)(\Omega, 0) = \vartheta_{S,P} \begin{bmatrix} \alpha \\ \beta \end{bmatrix}(\Omega). \tag{2.6.10}$$

By Theorem 25,

$$\begin{aligned} \vartheta_S \begin{bmatrix} \alpha \\ \beta \end{bmatrix}(\Omega, Z) = & \det(C\Omega + D)^{-\frac{m}{2}} \\ & \times e^{-\pi i \sigma(Z(C\Omega+D)^{-1}C\, {}^tZS^{-1})} \\ & \times \vartheta_S \begin{bmatrix} \alpha \\ \beta \end{bmatrix}((A\Omega + B)(C\Omega + D)^{-1}, Z(C\Omega + D)^{-1}). \end{aligned} \tag{2.6.11}$$

If we apply the differential operator $(2\pi i)^{-k}P(\partial)$ to both sides of Formula (2.6.11) and put $Z = 0$, according to Formula (2.6.10), Lemma 40 and Lemma 41, we obtain

$$\begin{aligned} & \vartheta_{S,P} \begin{bmatrix} \alpha \\ \beta \end{bmatrix}(\Omega) \\ = & (2\pi i)^{-k} \det(C\Omega + D)^{-\frac{m}{2}} \\ & \times \left[P(\partial)\vartheta_S \begin{bmatrix} \alpha \\ \beta \end{bmatrix}((A\Omega + B)(C\Omega + D)^{-1}, Z(C\Omega + D)^{-1}) \right]_{Z=0} \\ = & (2\pi i)^{-k} \det(C\Omega + D)^{-\frac{m}{2}} \left[\widetilde{P}(\partial)\vartheta_S \begin{bmatrix} \alpha \\ \beta \end{bmatrix}((A\Omega + B)(C\Omega + D)^{-1}, Z) \right]_{Z=0} \\ = & \det(C\Omega + D)^{-\frac{m}{2}} \vartheta_{S,\widetilde{P}} \begin{bmatrix} \alpha \\ \beta \end{bmatrix}((A\Omega + B)(C\Omega + D)^{-1}, 0) \\ = & \det(C\Omega + D)^{-\frac{m}{2}} \vartheta_{S,\widetilde{P}} \begin{bmatrix} \alpha \\ \beta \end{bmatrix}((A\Omega + B)(C\Omega + D)^{-1}), \end{aligned}$$

where $\widetilde{P}(Z) = P(Z(C\Omega + D))$. We note that we used Formula (2.6.6) and Lemma 40 in the first equality, and Lemma 41 in the second equality. In the third equality we used the fact that $\widetilde{P}$ is homogeneous of degree k. Consequently we complete the proof. □

Definition 16. Let (ρ, V_ρ) be a finite dimensional rational representation of $GL(n, \mathbb{C})$. A vector valued function $f : \mathbb{H}_n \longrightarrow V_\rho$ is called a *modular form* with respect to ρ if it is a holomorphic function on $\mathbb{H}_n$ such that

$$f((A\Omega + B)(C\Omega + D)^{-1}) = \rho(C\Omega + D)f(\Omega), \quad \Omega \in \mathbb{H}_n$$

for all $\begin{pmatrix} A & B \\ C & D \end{pmatrix}$ in a suitable congruence subgroup of Γ_n.

We recall that $\mathfrak{H}(S)$ denotes the space of all pluriharmonic polynomials in $\mathfrak{P}_{m,n}$ with respect to S. Let W be some $GL(n,\mathbb{C})$-stable subspace of $\mathfrak{H}(S)$. We define the W^*-valued function $\vartheta_W \begin{bmatrix} \alpha \\ \beta \end{bmatrix} : \mathbb{H}_n \longrightarrow W^*$ by

$$\left(\vartheta_W \begin{bmatrix} \alpha \\ \beta \end{bmatrix}(\Omega)\right)(P) := \vartheta_{S,P} \begin{bmatrix} \alpha \\ \beta \end{bmatrix}(\Omega) \tag{2.6.12}$$

for all $\Omega \in \mathbb{H}_n$ and $P \in W \subset \mathfrak{H}(S)$. Here W^* denotes the dual space of W.

Now we introduce the homogeneous line bundle $\mathcal{L}^{\frac{1}{2}}$ over $\mathbb{H}_n$. First of all we consider the double covering $\widetilde{GL(n,\mathbb{C})}$ of $GL(n,\mathbb{C})$ defined by

$$\widetilde{GL(n,\mathbb{C})} = \{ (g,\alpha) \mid \alpha^2 = \det(g), \ g \in GL(n,\mathbb{C}), \ \alpha \in \mathbb{C}^* \}$$

equipped with the multiplication

$$(g_1,\alpha_1)(g_2,\alpha_2) = (g_1 g_2, \alpha_1\alpha_2), \quad g_1, g_2 \in GL(n,\mathbb{C}), \ \alpha_1,\alpha_2 \in \mathbb{C}^*.$$

Let ρ be a one-dimensional representation of $GL(n,\mathbb{C})$ defined by

$$\rho(g,\alpha) = \alpha = (\det(g))^{\frac{1}{2}}, \quad g \in GL(n,\mathbb{C}), \ \alpha \in \mathbb{C}^*.$$

Then ρ yields the homogeneous line bundle on $\mathbb{H}_n$, denoted by $\mathcal{L}^{\frac{1}{2}}$. The complex manifold

$$\mathcal{L}^{\frac{1}{2}} = \mathbb{H}_n \times \mathbb{C}$$

is a holomorphic line bundle over $\mathbb{H}_n$ with the action of the metaplectic group $Mp(n,\mathbb{R})$ given by

$$\widetilde{M} \cdot (\Omega, z) = \left((A\Omega + B)(C\Omega + D)^{-1}, \det(C\Omega + D)^{1/2}\, z\right), \quad \widetilde{M} \in Mp(n,\mathbb{R}),$$

where $\begin{pmatrix} A & B \\ C & D \end{pmatrix} \in Sp(n,\mathbb{R})$ is the image of $\widetilde{M}$ under the surjective homomorphism of $Mp(n,\mathbb{R})$ onto $Sp(n,\mathbb{R})$. For a positive integer k, we define

$$\mathcal{L}^{\frac{k}{2}} = \left(\mathcal{L}^{\frac{1}{2}}\right)^{\otimes k} = \mathcal{L}^{\frac{1}{2}} \otimes \cdots \otimes \mathcal{L}^{\frac{1}{2}} \quad (k\text{-times}).$$

Let τ be the representation of $GL(n,\mathbb{C})$ on W defined by

$$\big(\tau(g)P\big)(Z) := P(Zg), \quad g \in GL(n,\mathbb{C}), \ P \in W, \ Z \in \mathbb{C}^{(m,n)}.$$

We observe that if $\widetilde{P}$ is a homogeneous pluriharmonic polynomial given by Theorem 16.9, then $\widetilde{P} = \tau(C\Omega + D)P$. Let τ^* be the contragredient of τ. That is,

$$\big(\tau^*(g)\ell\big)(P) = \ell\big(\tau(g)^{-1}P\big), \quad g \in GL(n,\mathbb{C}), \ \ell \in W^*, \ P \in W.$$

Theorem 27. *Let α and β as above. Then the function $\vartheta_W\begin{bmatrix}\alpha\\ \beta\end{bmatrix}(\Omega)$ defined in (2.6.12) is a modular form with values in $W^*\otimes\mathcal{L}^{\frac{m}{2}}$ with respect to the representation $\tau^*\otimes\det^{\frac{m}{2}}$ for a suitable congruence subgroup Γ. For any W and Ω, it is non-zero for suitable α and β.*

Proof. By Theorem 26, for all $P\in W\subset\mathfrak{H}(S)$ and for all $\begin{pmatrix}A & B\\ C & D\end{pmatrix}$ in a suitable congruence subgroup Γ of Γ_n, we get

$$
\begin{aligned}
&\left(\vartheta_W\begin{bmatrix}\alpha\\ \beta\end{bmatrix}(\Omega)\right)(P)\\
&=\det(C\Omega+D)^{-\frac{m}{2}}\cdot\vartheta_{S,\widetilde{P}}\begin{bmatrix}\alpha\\ \beta\end{bmatrix}\left((A\Omega+B)(C\Omega+D)^{-1}\right)\\
&=\det(C\Omega+D)^{-\frac{m}{2}}\cdot\left(\vartheta_W\begin{bmatrix}\alpha\\ \beta\end{bmatrix}\left((A\Omega+B)(C\Omega+D)^{-1}\right)\right)(\widetilde{P})\\
&=\det(C\Omega+D)^{-\frac{m}{2}}\cdot\left(\vartheta_W\begin{bmatrix}\alpha\\ \beta\end{bmatrix}\left((A\Omega+B)(C\Omega+D)^{-1}\right)\right)(\tau(C\Omega+D)P)\\
&=\det(C\Omega+D)^{-\frac{m}{2}}\cdot\left(\tau^*(C\Omega+D)^{-1}\vartheta_W\begin{bmatrix}\alpha\\ \beta\end{bmatrix}\left((A\Omega+B)(C\Omega+D)^{-1}\right)\right)(P),
\end{aligned}
$$

where $\widetilde{P}$ is a homogeneous pluriharmonic polynomial defined by $\widetilde{P}(Z)=P(Z(C\Omega+D))$. Therefore

$$
\vartheta_W\begin{bmatrix}\alpha\\ \beta\end{bmatrix}\left((A\Omega+B)(C\Omega+D)^{-1}\right)=\det(C\Omega+D)^{\frac{m}{2}}\cdot\tau^*(C\Omega+D)\vartheta_W\begin{bmatrix}\alpha\\ \beta\end{bmatrix}(\Omega).
$$

Hence $\vartheta_W\begin{bmatrix}\alpha\\ \beta\end{bmatrix}(\Omega)$ is a modular form on $\mathbb{H}_n$ with values in $W^*\otimes\mathcal{L}^{\frac{m}{2}}$ with respect to a suitable congruence subgroup Γ of Γ_n. □

Remark 7. Using Theorem 27, we can prove that for all $n\geq 2$ and $1\leq r\leq n-1$, there are congruence subgroups $\Gamma\subset\Gamma_n$ and Γ-invariant non-vanishing holomorphic k-forms on $\mathbb{H}_n$, where $k=\frac{n(n+1)}{2}-\frac{r(r+1)}{2}$. The proof can be found in [28]. This fact was proved by Freitag and Stillman.

Definition 17. Let (ρ,V_ρ) be a finite dimensional rational representation of $GL(n,\mathbb{C})$. A **pluriharmonic form** with respect to ρ is a polynomial P from $\mathbb{C}^{(m,n)}$ to V_ρ if it satisfies the following conditions:

$$
\sum_{k=1}^{m}\frac{\partial^2 P}{\partial z_{ki}\partial z_{kj}}=0\qquad\text{for all } i,j=1,2,\cdots,n \tag{2.6.13}
$$

and

$$
P(ZA)=\rho({}^tA)P(Z)\qquad\text{for all } A\in GL(n,\mathbb{C}). \tag{2.6.14}
$$

We denote by $\mathfrak{H}_{m,n}(\rho)$ the space of all pluriharmonic forms with respect to ρ.

Freitag proved the following.

Theorem 28. *Let S be a positive definite even unimodular matrix of degree m and let (ρ, V_ρ) be a finite dimensional rational representation of $GL(n,\mathbb{C})$. Let $P \in \mathfrak{H}_{m,n}(\rho)$ be a pluriharmonic form with respect to ρ. Then the theta series*

$$\Theta_{S,P}(\Omega) := \sum_{N \in \mathbb{Z}^{(m,n)}} P(S^{1/2}N)\, e^{\pi\, i\, \sigma({}^tNSN\Omega)} \tag{2.6.15}$$

is a modular form with respect to the representation ρ_* of $GL(n,\mathbb{C})$ defined by

$$\rho_*(A) \;=\; \rho(A)\,(\det A)^{\frac{m}{2}}, \qquad A \in GL(n,\mathbb{C})$$

for the the Siegel modular group Γ_n.

Proof. We will omit the proof. The proof can be found in [7]. □

2.7 Relation between Theta Series and the Weil Representation

Let (π, V_π) be a unitary projective representation of $Sp(n,\mathbb{R})$ on the representation space V_π. We assume that (π, V_π) satisfies the following conditions (A) and (B):

(A) There exists a vector valued map

$$\mathscr{F} : \mathbb{H}_n \longrightarrow V_\pi, \qquad \Omega \mapsto \mathscr{F}_\Omega := \mathscr{F}(\Omega)$$

satisfying the following covariance relation

$$\pi(M)\mathscr{F}_\Omega = \psi(M)\, J(M,\Omega)^{-1}\, \mathscr{F}_{M\cdot\Omega} \tag{2.7.1}$$

for all $M \in Sp(n,\mathbb{R})$ and $\Omega \in \mathbb{H}_n$. Here ψ is a character of $Sp(n,\mathbb{R})$ and $J : Sp(n,\mathbb{R}) \times \mathbb{H}_n \longrightarrow GL(1,\mathbb{C})$ is a certain automorphic factor for $Sp(n,\mathbb{R})$ on $\mathbb{H}_n$.

(B) Let Γ be an arithmetic subgroup of the Siegel modular group Γ_n. There exists a linear functional $\theta : V_\pi \longrightarrow \mathbb{C}$ which is semi-invariant under the action of Γ, in other words, for all $\gamma \in \Gamma$ and $\Omega \in \mathbb{H}_n$,

$$\langle\, \pi^*(\gamma)\theta,\, \mathscr{F}_\Omega \,\rangle = \langle\, \theta, \pi(\gamma)^{-1}\mathscr{F}_\Omega \,\rangle = \chi(\gamma)\,\langle\, \theta,\, \mathscr{F}_\Omega \,\rangle, \tag{2.7.2}$$

where π^* is the contragredient of π and $\chi : \Gamma \longrightarrow \mathbb{C}_1^*$ is a unitary character of Γ.

Under the assumptions (A) and (B) on a unitary projective representation (π, V_π), we define the function Θ on $\mathbb{H}_n$ by

$$\Theta(\Omega) := \langle\, \theta, \mathscr{F}_\Omega \,\rangle = \theta\big(\mathscr{F}_\Omega\big), \qquad \Omega \in \mathbb{H}_n. \tag{2.7.3}$$

We now shall see that Θ is an automorphic form on $\mathbb{H}_n$ with respect to Γ for the automorphic factor J.

Lemma 43. *Let (π, V_π) be a unitary projective representation of $Sp(n,\mathbb{R})$ satisfying the above assumptions (A) and (B). Then the function Θ on $\mathbb{H}_n$ defined by (17.3) satisfies the following modular transformation behavior*

$$\Theta(\gamma\cdot\Omega) = \psi(\gamma)^{-1}\,\chi(\gamma)^{-1}\, J(\gamma,\Omega)\,\Theta(\Omega) \tag{2.7.4}$$

for all $\gamma \in \Gamma$ and $\Omega \in \mathbb{H}_n$.

Proof. For any $\gamma \in \Gamma$ and $\Omega \in \mathbb{H}_n$, according to the assumptions (2.7.1) and (2.7.2), we obtain

$$\begin{aligned}\Theta(\gamma\cdot\Omega) &= \langle\,\theta,\mathscr{F}_{\gamma\cdot\Omega}\rangle\\ &= \langle\,\theta,\psi(\gamma)^{-1}\,J(\gamma,\Omega)\,\pi(\gamma)\mathscr{F}_{\Omega}\,\rangle\\ &= \psi(\gamma)^{-1}J(\gamma,\Omega)\,\langle\,\theta,\pi(\gamma)\mathscr{F}_{\Omega}\,\rangle\\ &= \psi(\gamma)^{-1}\,\chi(\gamma)^{-1}\,J(\gamma,\Omega)\,\langle\,\theta,\mathscr{F}_{\Omega}\,\rangle\\ &= \psi(\gamma)^{-1}\,\chi(\gamma)^{-1}\,J(\gamma,\Omega)\,\Theta(\Omega).\end{aligned}$$

□

Now for a positive definite real symmetric matrix $\mathcal{M}$ of degree m, we define the holomorphic function $\Theta_{\mathcal{M}}:\mathbb{H}_n\longrightarrow\mathbb{C}$ by

$$\Theta_{\mathcal{M}}(\Omega):=\sum_{\xi\in\mathbb{Z}^{(m,n)}}e^{2\pi i\sigma(\mathcal{M}\xi\,\Omega\,{}^t\xi)},\qquad \Omega\in\mathbb{H}_n. \tag{2.7.5}$$

Theorem 29. *Let $2\,\mathcal{M}$ be a symmetric positive definite, unimodular even integral matrix of degree m. Then for any $\gamma\in\Gamma_n$, the function $\Theta_{\mathcal{M}}$ satisfies the functional equation*

$$\Theta_{\mathcal{M}}(\gamma\cdot\Omega)=\rho_{\mathcal{M}}(\gamma)\,J_m(\gamma,\Omega)\Theta_{\mathcal{M}}(\Omega),\qquad \Omega\in\mathbb{H}_n, \tag{2.7.6}$$

where $\rho_{\mathcal{M}}$ is a character of Γ with $|\rho_{\mathcal{M}}(\gamma)|^8=1$ for all $\gamma\in\Gamma_n$ and $J_m: Sp(n,\mathbb{R})\times\mathbb{H}_n\longrightarrow\mathbb{C}_1^*$ is the automorphic factor for $Sp(n,\mathbb{R})$ on $\mathbb{H}_n$ defined by the formula (2.4.2) in Section 2.4.

Proof. For an element $\gamma=\begin{pmatrix}A&B\\C&D\end{pmatrix}\in\Gamma_n$ and $\Omega\in\mathbb{H}_n$, we put

$$\Omega_*=\gamma\cdot\Omega=(A\Omega+B)(C\Omega+D)^{-1}.$$

We define the linear functional ϑ on $L^2\big(\mathbb{R}^{(m,n)}\big)$ by

$$\vartheta(f)=\langle\vartheta,f\rangle:=\sum_{\xi\in\mathbb{Z}^{(m,n)}}f(\xi),\qquad f\in L^2\big(\mathbb{R}^{(m,n)}\big).$$

We note that $\Theta_{\mathcal{M}}(\Omega)=\vartheta\big(\mathscr{F}_{\Omega}^{(\mathcal{M})}\big)$. Since $\mathscr{F}^{(\mathcal{M})}$ is a covariant map for the Weil representation $\omega_{\mathcal{M}}$ with respect to the automorphic factor J_m by Theorem 16, according to Lemma 43, it suffices to prove that ϑ is semi-invariant for $\omega_{\mathcal{M}}$ under the action of Γ_n, in other words, ϑ satisfies the following semi-invariance relation

$$\Big\langle\,\vartheta,R_{\mathcal{M}}(\gamma)\mathscr{F}_{\Omega}^{(\mathcal{M})}\,\Big\rangle=\rho_{\mathcal{M}}(\gamma)^{-1}\Big\langle\,\vartheta,\mathscr{F}_{\Omega}^{(\mathcal{M})}\,\Big\rangle \tag{2.7.7}$$

for all $\gamma\in\Gamma_n$ and $\Omega\in\mathbb{H}_n$.

We see that the following elements

$$t_\beta = \begin{pmatrix} I_n & \beta \\ 0 & I_n \end{pmatrix} \text{ with any } \beta = {}^t\beta \in \mathbb{Z}^{(n,n)},$$

$$d_\alpha = \begin{pmatrix} {}^t\alpha & 0 \\ 0 & \alpha^{-1} \end{pmatrix} \text{ with any } \alpha \in GL(n,\mathbb{Z}),$$

$$\sigma_n = \begin{pmatrix} 0 & -I_n \\ I_n & 0 \end{pmatrix}$$

generate the Siegel modular group Γ_n. Therefore it suffices to prove the semi-invariance relation (2.7.7) for the above generators of Γ_n.

Case I. $\gamma = t_\beta$ with $\beta = {}^t\beta \in \mathbb{Z}^{(n,n)}$.

In this case, we have

$$\Omega_* = \Omega + \beta \qquad \text{and} \qquad J_m(\gamma, \Omega) = 1.$$

According to the covariance relation (2.4.4) in Section 2.4, we obtain

$$\begin{aligned}
&\big\langle\, \vartheta, R_{\mathcal{M}}(\gamma)\mathscr{F}_\Omega^{(\mathcal{M})} \,\big\rangle \\
&= \big\langle\, \vartheta, J_m(\gamma,\Omega)^{-1}\mathscr{F}_{\gamma\cdot\Omega}^{(\mathcal{M})} \,\big\rangle \\
&= \big\langle\, \vartheta, \mathscr{F}_{\Omega+\beta}^{(\mathcal{M})} \,\big\rangle \\
&= \sum_{\xi\in\mathbb{Z}^{(m,n)}} \mathscr{F}_{\Omega+\beta}^{(\mathcal{M})}(\xi) \\
&= \sum_{\xi\in\mathbb{Z}^{(m,n)}} e^{2\pi i\sigma(\mathcal{M}\,\xi\,(\Omega+\beta)\,{}^t\xi)} \\
&= \sum_{\xi\in\mathbb{Z}^{(m,n)}} e^{2\pi i\sigma(\mathcal{M}\,\xi\,\Omega\,{}^t\xi)}\cdot e^{2\pi i\sigma(\mathcal{M}\,\xi\,\beta\,{}^t\xi)} \\
&= \sum_{\xi\in\mathbb{Z}^{(m,n)}} e^{2\pi i\sigma(\mathcal{M}\,\xi\,\Omega\,{}^t\xi)} \\
&= \big\langle\, \vartheta, \mathscr{F}_\Omega^{(\mathcal{M})} \,\big\rangle.
\end{aligned}$$

Here we used the fact that $2\,\sigma(\mathcal{M}\,\xi\,\beta\,{}^t\xi)$ is an even integer because $2\,\mathcal{M}$ is even integral. We put $\rho_{\mathcal{M}}(\gamma) = \rho_{\mathcal{M}}(t_\beta) = 1$ for all $\beta = {}^t\beta \in \mathbb{Z}^{(n,n)}$. Therefore ϑ satisfies the semi-invariance relation (2.7.7) in the case $\gamma = t_\beta$ with $\beta = {}^t\beta \in \mathbb{Z}^{(n,n)}$.

Case II. $\gamma = d_\alpha$ with $\alpha \in GL(n,\mathbb{Z})$.

In this case, we have

$$\Omega_* = {}^t\alpha\,\Omega\,\alpha \qquad \text{and} \qquad J_m(d_\alpha, \Omega) = (\det\alpha)^{-\frac{m}{2}}.$$

According to the covariance relation (2.4.4) in Section 2.4, we obtain

$$\begin{aligned}
&\left\langle \vartheta, R_{\mathcal{M}}(\gamma)\mathscr{F}_{\Omega}^{(\mathcal{M})} \right\rangle \text{ with } \gamma = d_\alpha \\
&= \left\langle \vartheta, J_m(\gamma,\Omega)^{-1}\mathscr{F}_{\gamma\cdot\Omega}^{(\mathcal{M})} \right\rangle \\
&= (\det\alpha)^{\frac{m}{2}} \left\langle \vartheta, \mathscr{F}_{{}^t\alpha\,\Omega\,\alpha}^{(\mathcal{M})} \right\rangle \\
&= (\det\alpha)^{\frac{m}{2}} \sum_{\xi\in\mathbb{Z}^{(m,n)}} \mathscr{F}_{{}^t\alpha\,\Omega\,\alpha}^{(\mathcal{M})}(\xi) \\
&= (\det\alpha)^{\frac{m}{2}} \sum_{\xi\in\mathbb{Z}^{(m,n)}} e^{2\pi i\sigma(\mathcal{M}\,\xi\,{}^t\alpha\,\Omega\,\alpha\,{}^t\xi)} \\
&= (\det\alpha)^{\frac{m}{2}} \sum_{\xi\in\mathbb{Z}^{(m,n)}} e^{2\pi i\sigma\{\mathcal{M}((\xi\,{}^t\alpha)\,\Omega\,{}^t(\xi\,{}^t\alpha))\}} \\
&= (\det\alpha)^{\frac{m}{2}} \left\langle \vartheta, \mathscr{F}_{\Omega}^{(\mathcal{M})} \right\rangle.
\end{aligned}$$

Here we put $\rho_{\mathcal{M}}(d_\alpha) = (\det\alpha)^{-\frac{m}{2}}$. Therefore ϑ satisfies the semi-invariance relation (2.7.7) in the case $\gamma = d_\alpha$ with $\alpha \in GL(n,\mathbb{Z})$.

Case III. $\gamma = \sigma_n = \begin{pmatrix} 0 & -I_n \\ I_n & 0 \end{pmatrix}$.

In this case, we have

$$\Omega_* = -\Omega^{-1} \qquad \text{and} \qquad J_m(\sigma_n,\Omega) = \left(\det\Omega\right)^{\frac{m}{2}}.$$

In the process of the proof of Theorem 16, using Lemma 34, we already showed that

$$\begin{aligned}
&\int_{\mathbb{R}^{(m,n)}} e^{2\pi i\sigma(\mathcal{M}(y\,\Omega\,{}^ty + 2\,y\,{}^tx))}dy \qquad (2.7.8)\\
&= \left(\det\mathcal{M}\right)^{-\frac{n}{2}} \left(\det\frac{2\,\Omega}{i}\right)^{-\frac{m}{2}} e^{-2\pi i\sigma(\mathcal{M}\,x\,\Omega^{-1}\,{}^tx)}.
\end{aligned}$$

By Formula (2.7.8), we obtain

$$\begin{aligned}
\widehat{\mathscr{F}_{\Omega}^{(\mathcal{M})}}(2\,\mathcal{M}x) &= \int_{\mathbb{R}^{(m,n)}} \mathscr{F}_{\Omega}^{(\mathcal{M})}(y)\, e^{-2\pi i\sigma(y\,{}^t(2\,\mathcal{M}\,x))}\,dy \\
&= \int_{\mathbb{R}^{(m,n)}} e^{2\pi i\sigma(\mathcal{M}\,y\,\Omega\,{}^ty)}\cdot e^{-4\pi i\sigma(\mathcal{M}\,y\,{}^tx)}\,dy \\
&= \int_{\mathbb{R}^{(m,n)}} e^{2\pi i\sigma\{\mathcal{M}(y\,\Omega\,{}^ty + 2\,y\,{}^t(-x))\}}\,dy \\
&= (\det\mathcal{M})^{-\frac{n}{2}} \left(\det\frac{2\,\Omega}{i}\right)^{-\frac{m}{2}} e^{-2\pi i\sigma(\mathcal{M}\,(-x)\,\Omega^{-1}\,{}^t(-x))} \\
&= (\det\mathcal{M})^{-\frac{n}{2}} \left(\det\frac{2\,\Omega}{i}\right)^{-\frac{m}{2}} e^{-2\pi i\sigma(\mathcal{M}\,x\,\Omega^{-1}\,{}^tx)}.
\end{aligned}$$

Thus we obtain

$$\widehat{\mathscr{F}_{\Omega}^{(\mathcal{M})}}(2\,\mathcal{M}x) = \big(\det\mathcal{M}\big)^{-\frac{n}{2}}\left(\det\frac{2\,\Omega}{i}\right)^{-\frac{m}{2}} e^{-2\,\pi\,i\,\sigma(\mathcal{M}\,x\,\Omega^{-1}\,{}^t x)}, \tag{2.7.9}$$

where $\widehat{f}$ is the Fourier transform of f defined by

$$\widehat{f}(x) = \int_{\mathbb{R}^{(m,n)}} f(y)\,e^{-2\,\pi\,i\,\sigma(y\,{}^t x)}\,dy, \quad x\in\mathbb{R}^{(m,n)}.$$

We prove the Poisson summation formula in our setting.

Lemma 44. *Let f be an element in $L^2\big(\mathbb{R}^{(m,n)}\big)$. Then*

$$\sum_{\xi\in\mathbb{Z}^{(m,n)}}\widehat{f}(\xi) = \sum_{\xi\in\mathbb{Z}^{(m,n)}} f(\xi). \tag{2.7.10}$$

Proof. We define

$$h(x) = \sum_{\xi\in\mathbb{Z}^{(m,n)}} f(x+\xi), \qquad x\in\mathbb{R}^{(m,n)}. \tag{2.7.11}$$

We see that $h(x)$ is periodic in x_{ij} with period 1, where $x = (x_{ij})$ is a coordinate in $\mathbb{R}^{(m,n)}$. Thus $h(x)$ has the following Fourier series

$$h(x) = \sum_{\xi\in\mathbb{Z}^{(m,n)}} c_\xi\,e^{2\,\pi\,i\,\sigma(x\,{}^t\xi)}, \tag{2.7.12}$$

where

$$\begin{aligned} c_\xi &= \int_0^1\cdots\int_0^1 h(x)\,e^{-2\,\pi\,i\,\sigma(x\,{}^t\xi)}\,dx \\ &= \int_0^1\cdots\int_0^1 \sum_{\xi\in\mathbb{Z}^{(m,n)}} f(x+\xi)\,e^{-2\,\pi\,i\,\sigma(x\,{}^t\xi)}\,dx \\ &= \int_{\mathbb{R}^{(m,n)}} f(x)\,e^{-2\,\pi\,i\,\sigma(x\,{}^t\xi)}\,dx \;=\; \widehat{f}(\xi). \end{aligned}$$

Here we interchanged summation and integration, and made a change of variables replacing $x+\xi$ by x to obtain the above equality.
By the definition (2.7.11), we have

$$h(0) = \sum_{\xi\in\mathbb{Z}^{(m,n)}} f(\xi).$$

On the other hand, from Formula (2.7.12), we get

$$h(0) = \sum_{\xi \in \mathbb{Z}^{(m,n)}} c_\xi = \sum_{\xi \in \mathbb{Z}^{(m,n)}} \widehat{f}(\xi).$$

Therefore we obtain the Poisson summation formula (2.7.10). □

According to the covariance relation (2.4.4) in Section 2.4, Formula (2.7.9) and Poisson summation formula, we obtain

$$\begin{aligned}
&\big\langle \vartheta, R_{\mathcal{M}}(\gamma)\mathscr{F}_\Omega^{(\mathcal{M})} \big\rangle \quad \text{with } \gamma = \sigma_n \\
= &\big\langle \vartheta, J_m(\gamma,\Omega)^{-1}\mathscr{F}_{\gamma\cdot\Omega}^{(\mathcal{M})} \big\rangle \\
= &J_m(\gamma,\Omega)^{-1}\big\langle \vartheta, \mathscr{F}_{-\Omega^{-1}}^{(\mathcal{M})} \big\rangle \\
= &(\det \Omega)^{-\frac{m}{2}} \sum_{\xi \in \mathbb{Z}^{(m,n)}} e^{-2\pi i \sigma(\mathcal{M}\xi\Omega^{-1}\,{}^t\xi)} \\
= &(\det \Omega)^{-\frac{m}{2}} \big(\det \mathcal{M}\big)^{\frac{n}{2}} \left(\det \frac{2\Omega}{i}\right)^{\frac{m}{2}} \sum_{\xi \in \mathbb{Z}^{(m,n)}} \widehat{\mathscr{F}_\Omega^{(\mathcal{M})}}(2\mathcal{M}\xi) \\
&\quad (\text{by Formula (2.7.9)}) \\
= &\big(\det 2\mathcal{M}\big)^{\frac{n}{2}} \left(\det \frac{I_n}{i}\right)^{\frac{m}{2}} \sum_{\xi \in \mathbb{Z}^{(m,n)}} \widehat{\mathscr{F}_\Omega^{(\mathcal{M})}}(\xi) \\
&\quad (\text{because } 2\mathcal{M} \text{ is unimodular})) \\
= &\left(\det \frac{I_n}{i}\right)^{\frac{m}{2}} \sum_{\xi \in \mathbb{Z}^{(m,n)}} \mathscr{F}_\Omega^{(\mathcal{M})}(\xi) \quad (\text{by Poisson summation formula}) \\
= &(-i)^{\frac{mn}{2}} \big\langle \vartheta, \mathscr{F}_\Omega^{(\mathcal{M})} \big\rangle \\
= &(-i)^{\frac{mn}{2}} \big\langle \vartheta, \mathscr{F}_\Omega^{(\mathcal{M})} \big\rangle.
\end{aligned}$$

We put $\rho_{\mathcal{M}}(\sigma_n) = (-i)^{-\frac{mn}{2}}$. Therefore ϑ satisfies the semi-invariance relation (2.7.7) in the case $\gamma = \sigma_n$. The proof of Case III is completed. Since J_m is an automorphic factor for $Sp(n,\mathbb{R})$ on $\mathbb{H}_n$, we see that if the formula (2.7.6) holds for two elements γ_1, γ_2 in Γ, then it holds for $\gamma_1\gamma_2$. Finally we complete the proof of Theorem 29. □

Remark 8. For a symmetric positive definite integral matrix $\mathcal{M}$ such that $2\mathcal{M}$ is not unimodular even integral, we obtain a similar transformation formula like (2.7.6). If m is odd, $\Theta_{\mathcal{M}}(\Omega)$ is a modular form of a half-integral weight $\frac{m}{2}$ and index $\frac{\mathcal{M}}{2}$ with respect to a suitable arithmetic subgroup $\Gamma_{\Theta,\mathcal{M}}$ of Γ_n and a suitable character $\rho_{\mathcal{M}}$ of $\Gamma_{\Theta,\mathcal{M}}$.

2.8 Spectral Theory on the Abelian Variety

We recall the Jacobi group (cf. Section 2.3)

$$G^J = Sp(n,\mathbb{R}) \ltimes H_{\mathbb{R}}^{(n,m)}$$

which is the semidirect product of $Sp(n,\mathbb{R})$ and $H_{\mathbb{R}}^{(n,m)}$ endowed with the following multiplication law

$$\Big(M,(\lambda,\mu,\kappa)\Big)\Big(M',(\lambda',\mu',\kappa')\Big) = \Big(MM',(\widetilde{\lambda}+\lambda',\widetilde{\mu}+\mu',\kappa+\kappa'+\widetilde{\lambda}\,{}^t\mu'-\widetilde{\mu}\,{}^t\lambda')\Big)$$

with $M, M' \in Sp(n,\mathbb{R}), (\lambda,\mu,\kappa),\ (\lambda',\mu',\kappa') \in H_{\mathbb{R}}^{(n,m)}$ and $(\widetilde{\lambda},\widetilde{\mu}) = (\lambda,\mu)M'$. Then G^J acts on $\mathbb{H}_n \times \mathbb{C}^{(m,n)}$ transitively by

$$\begin{aligned} &\big(M,(\lambda,\mu,\kappa)\big)\cdot(\Omega,Z) \\ &= \big((A\Omega+B)(C\Omega+D)^{-1},(Z+\lambda\Omega+\mu)(C\Omega+D)^{-1}\big), \end{aligned} \tag{2.8.1}$$

where $M = \begin{pmatrix} A & B \\ C & D \end{pmatrix} \in Sp(n,\mathbb{R}),\ (\lambda,\mu,\kappa) \in H_{\mathbb{R}}^{(n,m)}$ and $(\Omega,Z) \in \mathbb{H}_n \times \mathbb{C}^{(m,n)}$. We note that the Jacobi group G^J is *not* a reductive Lie group and also that the space $\mathbb{H}_n \times \mathbb{C}^{(m,n)}$ is not a symmetric space. We refer to [50]-[56] and [58] about automorphic forms on G^J and topics related to the content of this book.

From now on, for brevity, we write

$$\mathbb{H}_{n,m} := \mathbb{H}_n \times \mathbb{C}^{(m,n)}.$$

$\mathbb{H}_{n,m}$ is called the Siegel-Jacobi space of degree n and index m.

We let

$$\Gamma_{n,m} := \Gamma_n \ltimes H_{\mathbb{Z}}^{(n,m)}$$

be the discrete subgroup of G^J, where

$$H_{\mathbb{Z}}^{(n,m)} = \Big\{ (\lambda,\mu,\kappa) \in H_{\mathbb{R}}^{(n,m)} \mid \lambda,\mu \in \mathbb{Z}^{(m,n)},\ \ \kappa \in \mathbb{Z}^{(m,m)} \Big\}.$$

Let E_{kj} be the $m \times n$ matrix with entry 1 where the k-th row and the j-th colume meet, and all other entries 0. For an element $\Omega \in \mathbb{H}_n$, we set for brevity

$$F_{kj}(\Omega) := E_{kj}\Omega, \qquad 1 \le k \le m,\ 1 \le j \le n. \tag{2.8.2}$$

For each $\Omega \in \mathcal{F}_n$, we define a subset P_Ω of $\mathbb{C}^{(m,n)}$ by

$$P_\Omega = \left\{ \sum_{k=1}^{m}\sum_{j=1}^{n} \lambda_{kj}E_{kj} + \sum_{k=1}^{m}\sum_{j=1}^{n} \mu_{kj}F_{kj}(\Omega) \,\Big|\, 0 \le \lambda_{kj}, \mu_{kj} \le 1 \right\}.$$

For each $\Omega \in \mathcal{F}_n$, we define the subset D_Ω of $\mathbb{H}_{n,m}$ by

$$D_\Omega := \{ (\Omega, Z) \in \mathbb{H}_{n,m} \mid Z \in P_\Omega \}.$$

We define

$$\mathcal{F}_{n,m} := \cup_{\Omega \in \mathcal{F}_n} D_\Omega.$$

Theorem 30. *$\mathcal{F}_{n,m}$ is a fundamental domain for $\Gamma_{n,m}\backslash\mathbb{H}_{n,m}$.*

Proof. Let $(\tilde{\Omega}, \tilde{Z})$ be an arbitrary element of $\mathbb{H}_{n,m}$. We must find an element (Ω, Z) of $\mathcal{F}_{n,m}$ and an element $\gamma^J = (\gamma, (\lambda, \mu; \kappa)) \in \Gamma_{n,m}$ with $\gamma \in \Gamma_n$ such that $\gamma^J \cdot (\Omega, Z) = (\tilde{\Omega}, \tilde{Z})$. Since $\mathcal{F}_n$ is a fundamental domain for $\Gamma_n\backslash\mathbb{H}_n$, there exists an element γ of Γ_n and an element Ω of $\mathcal{F}_n$ such that $\gamma \cdot \Omega = \tilde{\Omega}$. Here Ω is unique up to the boundary of $\mathcal{F}_n$.

We write

$$\gamma = \begin{pmatrix} A & B \\ C & D \end{pmatrix} \in \Gamma_n.$$

It is easy to see that we can find $\lambda, \mu \in \mathbb{Z}^{(m,n)}$ and $Z \in P_\Omega$ satisfying the equation

$$Z + \lambda\Omega + \mu = \tilde{Z}(C\Omega + D).$$

If we take $\gamma^J = (\gamma, (\lambda, \mu; 0)) \in \Gamma_{n,m}$, we see that $\gamma^J \cdot (\Omega, Z) = (\tilde{\Omega}, \tilde{Z})$. Therefore we obtain

$$\mathbb{H}_{n,m} = \cup_{\gamma^J \in \Gamma_{n,m}} \gamma^J \cdot \mathcal{F}_{n,m}.$$

Let (Ω, Z) and $\gamma^J \cdot (\Omega, Z)$ be two elements of $\mathcal{F}_{n,m}$ with $\gamma^J = (\gamma, (\lambda, \mu; \kappa)) \in \Gamma_{n,m}$. Then both Ω and $\gamma \cdot \Omega$ lie in $\mathcal{F}_n$. Therefore both of them either lie in the boundary of $\mathcal{F}_n$ or $\gamma = \pm I_{2n}$. In the case that both Ω and $\gamma \cdot \Omega$ lie in the boundary of $\mathcal{F}_n$, both (Ω, Z) and $\gamma^J \cdot (\Omega, Z)$ lie in the boundary of $\mathcal{F}_{n,m}$. If $\gamma = \pm I_{2n}$, we have

$$Z \in P_\Omega \quad \text{and} \quad \pm (Z + \lambda\Omega + \mu) \in P_\Omega, \quad \lambda, \mu \in \mathbb{Z}^{(m,n)}. \tag{2.8.3}$$

From the definition of P_Ω and (2.8.3), we see that either $\lambda = \mu = 0$, $\gamma \neq -I_{2n}$ or both Z and $\pm(Z + \lambda\Omega + \mu)$ lie on the boundary of the parallelepiped P_Ω. Hence either both(Ω, Z) and $\gamma^J \cdot (\Omega, Z)$ lie in the boundary of $\mathcal{F}_{n,m}$ or $\gamma^J = (I_{2n}, (0, 0; \kappa)) \in \Gamma_{n,m}$. Consequently $\mathcal{F}_{n,m}$ is a fundamental domain for $\Gamma_{n,m}\backslash\mathbb{H}_{n,m}$. □

For a coordinate $(\Omega, Z) \in \mathbb{H}_{n,m}$ with $\Omega = (\omega_{\mu\nu}) \in \mathbb{H}_n$ and $Z = (z_{kl}) \in \mathbb{C}^{(m,n)}$, we put

$$\begin{aligned}
\Omega &= X + iY, \quad X = (x_{\mu\nu}), \quad Y = (y_{\mu\nu}) \text{ real},\\
Z &= U + iV, \quad U = (u_{kl}), \quad V = (v_{kl}) \text{ real},\\
d\Omega &= (d\omega_{\mu\nu}), \quad dX = (dx_{\mu\nu}), \quad dY = (dy_{\mu\nu}),\\
dZ &= (dz_{kl}), \quad dU = (du_{kl}), \quad dV = (dv_{kl}),\\
d\overline{\Omega} &= (d\overline{\omega}_{\mu\nu}), \quad d\overline{Z} = (d\bar{z}_{kl}),
\end{aligned}$$

$$\frac{\partial}{\partial\Omega} = \left(\frac{1+\delta_{\mu\nu}}{2}\frac{\partial}{\partial\omega_{\mu\nu}}\right), \quad \frac{\partial}{\partial\overline{\Omega}} = \left(\frac{1+\delta_{\mu\nu}}{2}\frac{\partial}{\partial\overline{\omega}_{\mu\nu}}\right),$$

$$\frac{\partial}{\partial Z} = \begin{pmatrix} \frac{\partial}{\partial z_{11}} & \cdots & \frac{\partial}{\partial z_{m1}} \\ \vdots & \ddots & \vdots \\ \frac{\partial}{\partial z_{1n}} & \cdots & \frac{\partial}{\partial z_{mn}} \end{pmatrix}, \quad \frac{\partial}{\partial \overline{Z}} = \begin{pmatrix} \frac{\partial}{\partial \overline{z}_{11}} & \cdots & \frac{\partial}{\partial \overline{z}_{m1}} \\ \vdots & \ddots & \vdots \\ \frac{\partial}{\partial \overline{z}_{1n}} & \cdots & \frac{\partial}{\partial \overline{z}_{mn}} \end{pmatrix}.$$

Remark 9. The following metric

$$\begin{aligned}
ds^2_{n,m} = &\,\sigma\left(Y^{-1}d\Omega\, Y^{-1}d\overline{\Omega}\right) + \sigma\left(Y^{-1}\,{}^tV\, V\, Y^{-1}d\Omega\, Y^{-1}d\overline{\Omega}\right)\\
&+ \sigma\left(Y^{-1}\,{}^t(dZ)\, d\overline{Z}\right)\\
&- \sigma\left(V\,Y^{-1}d\Omega\, Y^{-1}\,{}^t(d\overline{\Omega}) + V\,Y^{-1}d\overline{\Omega}\, Y^{-1}\,{}^t(dZ)\right)
\end{aligned}$$

is a Kähler metric on $\mathbb{H}_{n,m}$ which is invariant under the action (2.8.1) of the Jacobi group G^J. Its Laplacian is given by

$$\begin{aligned}
\Delta_{n,m} = &\,4\,\sigma\left(Y\,{}^t\!\left(Y\frac{\partial}{\partial\overline{\Omega}}\right)\frac{\partial}{\partial\Omega}\right) + 4\,\sigma\left(Y\frac{\partial}{\partial Z}\,{}^t\!\left(\frac{\partial}{\partial\overline{Z}}\right)\right)\\
&+ 4\,\sigma\left(VY^{-1}\,{}^tV\,{}^t\!\left(Y\frac{\partial}{\partial\overline{Z}}\right)\frac{\partial}{\partial Z}\right)\\
&+ 4\,\sigma\left(V\,{}^t\!\left(Y\frac{\partial}{\partial\overline{\Omega}}\right)\frac{\partial}{\partial Z}\right) + 4\sigma\left({}^tV\,{}^t\!\left(Y\frac{\partial}{\partial\overline{Z}}\right)\frac{\partial}{\partial\Omega}\right).
\end{aligned}$$

The following differential form

$$dv_{n,m} = (\det Y)^{-(n+m+1)}\,[dX]\wedge[dY]\wedge[dU]\wedge[dV]$$

is a G^J-invariant volume element on $\mathbb{H}_{n,m}$, where

$$[dX] = \bigwedge_{\mu\le\nu} dx_{\mu\nu}, \ [dY] = \bigwedge_{\mu\le\nu} dy_{\mu\nu}, \ [dU] = \bigwedge_{k,l} du_{kl} \ \text{ and } \ [dV] = \bigwedge_{k,l} dv_{kl}.$$

The point is that the invariant metric $ds^2_{n,m}$ and its Laplacian are beautifully expressed in terms of the *trace* form. The above facts were proved by the author in [48]. We also refer to [49] for the action of the Jacobi group G^J on the Siegel-Jacobi disk $\mathbb{D}_n \times \mathbb{C}^{(m,n)}$.

We fix two positive integers m and n throughout this section.

For an element $\Omega \in \mathbb{H}_n$, we set

$$L_\Omega := \mathbb{Z}^{(m,n)} + \mathbb{Z}^{(m,n)}\Omega$$

We use the notation (2.8.2). It follows from the positivity of $\operatorname{Im}\Omega$ that the elements $E_{kj},\, F_{kj}(\Omega)\ (1 \leq k \leq m,\ 1 \leq j \leq n)$ of L_Ω are linearly independent over $\mathbb{R}$. Therefore L_Ω is a lattice in $\mathbb{C}^{(m,n)}$ and the set

$$\{\, E_{kj},\, F_{kj}(\Omega) \mid 1 \leq k \leq m,\ 1 \leq j \leq n \}$$

forms an integral basis of L_Ω. We see easily that if Ω is an element of $\mathbb{H}_n$, the period matrix $\Omega_\flat := (I_n, \Omega)$ satisfies the Riemann conditions (RC.1) and (RC.2):

(RC.1) $\quad \Omega_\flat\, J_n\, {}^t\Omega_\flat = 0\,;$

(RC.2) $\quad -\frac{1}{i}\, \Omega_\flat\, J_n\, {}^t\overline{\Omega}_\flat > 0.$

Thus the complex torus $A_\Omega := \mathbb{C}^{(m,n)}/L_\Omega$ is an abelian variety. For more details on A_Ω, we refer to [14] and [27].

It might be interesting to investigate the spectral theory of the Laplacian $\Delta_{n,m}$ on a fundamental domain $\mathcal{F}_{n,m}$. But this work is very complicated and difficult at this moment. It may be that the first step is to develop the spectral theory of the Laplacian Δ_Ω on the abelian variety A_Ω. The second step will be to study the spectral theory of the Laplacian Δ_* (see (2.2.2) in Section 2.2) on the moduli space $\Gamma_n\backslash\mathbb{H}_n$ of principally polarized abelian varieties of dimension g. The final step would be to combine the above steps and more works to develop the spectral theory of the Lapalcian $\Delta_{n,m}$ on $\mathcal{F}_{n,m}$. In this section, we deal only with the spectral theory Δ_Ω on $L^2(A_\Omega)$.

We fix an element $\Omega = X + i\,Y$ of $\mathbb{H}_n$ with $X = \operatorname{Re}\Omega$ and $Y = \operatorname{Im}\Omega$. For a pair (A, B) with $A, B \in \mathbb{Z}^{(m,n)}$, we define the function $E_{\Omega;A,B} : \mathbb{C}^{(m,n)} \longrightarrow \mathbb{C}$ by

$$E_{\Omega;A,B}(Z) = e^{2\pi i\left(\sigma\,({}^t AU\,) + \sigma\,((B - AX)Y^{-1}\,{}^tV)\right)},$$

where $Z = U + iV$ is a variable in $\mathbb{C}^{(m,n)}$ with real U, V.

Lemma 45. *For any $A, B \in \mathbb{Z}^{(m,n)}$, the function $E_{\Omega;A,B}$ satisfies the following functional equation*

$$E_{\Omega;A,B}(Z + \lambda\Omega + \mu) = E_{\Omega;A,B}(Z), \quad Z \in \mathbb{C}^{(m,n)}$$

for all $\lambda, \mu \in \mathbb{Z}^{(m,n)}$. Thus $E_{\Omega;A,B}$ can be regarded as a function on A_Ω.

Proof. We write $\Omega = X + iY$ with real X, Y. For any $\lambda, \mu \in \mathbb{Z}^{(m,n)}$, we have

$$\begin{aligned}E_{\Omega;A,B}(Z+\lambda\Omega+\mu) &= E_{\Omega;A,B}((U+\lambda X+\mu)+i(V+\lambda Y)) \\ &= e^{2\pi i\{\,\sigma\,({}^tA(U+\lambda X+\mu))+\sigma\,((B-AX)Y^{-1}\,{}^t(V+\lambda Y))\,\}} \\ &= e^{2\pi i\{\,\sigma\,({}^tAU+{}^tA\lambda X+{}^tA\mu)+\sigma\,((B-AX)Y^{-1}\,{}^tV+B\,{}^t\lambda-AX\,{}^t\lambda)\}} \\ &= e^{2\pi i\{\,\sigma\,({}^tAU)+\sigma\,((B-AX)Y^{-1}\,{}^tV)\}} \\ &= E_{\Omega;A,B}(Z).\end{aligned}$$

Here we used the fact that ${}^tA\mu$ and $B\,{}^t\lambda$ are integral. □

Lemma 46. *The metric*

$$ds^2_\Omega = \sigma\left((\mathrm{Im}\,\Omega)^{-1}\,{}^t(dZ)\,d\overline{Z})\,\right)$$

is a Kähler metric on A_Ω invariant under the action (18.1) of $\Gamma^J = Sp(n,\mathbb{Z})\ltimes H_{\mathbb{Z}}^{(n,m)}$ on (Ω,Z) with Ω fixed. Its Laplacian Δ_Ω of ds^2_Ω is given by

$$\Delta_\Omega = \sigma\left((\mathrm{Im}\,\Omega)\,\frac{\partial}{\partial Z}\,{}^t\!\left(\frac{\partial}{\partial \overline{Z}}\right)\right).$$

Proof. Let $\tilde{\gamma}=(\gamma,(\lambda,\mu;\kappa))\in\Gamma^J$ with $\gamma=\begin{pmatrix}A & B\\ C & D\end{pmatrix}\in Sp(n,\mathbb{Z})$ and $(\tilde{\Omega},\tilde{Z}) = \tilde{\gamma}\cdot(\Omega,Z)$ with $\Omega\in\mathbb{H}_n$ fixed. Then according to [23, p. 33],

$$\mathrm{Im}\,\gamma\cdot\Omega = {}^t(C\overline{\Omega}+D)^{-1}\,\mathrm{Im}\,\Omega\,(C\Omega+D)^{-1}$$

and

$$d\tilde{Z} = dZ\,(C\Omega+D)^{-1}.$$

Therefore

$$\begin{aligned}&(\mathrm{Im}\,\tilde{\Omega})^{-1}\,{}^t(d\tilde{Z})\,d\overline{\tilde{Z}} \\ &= (C\overline{\Omega}+D)\,(\mathrm{Im}\,\Omega)^{-1}\,{}^t(C\Omega+D)\,{}^t(C\Omega+D)^{-1}\,{}^t(dZ)\,d\overline{Z}\,(C\overline{\Omega}+D)^{-1} \\ &= (C\overline{\Omega}+D)\,(\mathrm{Im}\,\Omega)^{-1}\,{}^t(dZ)\,d\overline{Z}\,(C\overline{\Omega}+D)^{-1}.\end{aligned}$$

The metric $ds_{iI_n} = \sigma(dZ\,{}^t(d\overline{Z}))$ at $Z=0$ is positive definite. Since G^J acts on $\mathbb{H}_{n,m}$ transitively, ds^2_Ω is a Riemannian metric for any $\Omega\in\mathbb{H}_n$. We note that the differential operator Δ_Ω is invariant under the action of Γ^J. In fact,

$$\frac{\partial}{\partial\tilde{Z}} = (C\Omega+D)\,\frac{\partial}{\partial Z}.$$

Hence if f is a differentiable function on A_Ω, then

$$\begin{aligned}
&\mathrm{Im}\,\tilde{\Omega}\,\frac{\partial}{\partial \tilde{Z}}{}^t\!\left(\frac{\partial f}{\partial \overline{\tilde{Z}}}\right) \\
&= {}^t(C\overline{\Omega}+D)^{-1}\,(\mathrm{Im}\,\Omega)\,(C\Omega+D)^{-1}(C\Omega+D)\,\frac{\partial}{\partial Z}{}^t\!\left((C\overline{\Omega}+D)\frac{\partial f}{\partial \overline{Z}}\right) \\
&= {}^t(C\overline{\Omega}+D)^{-1}\,\mathrm{Im}\,\Omega\,\frac{\partial}{\partial Z}{}^t\!\left(\frac{\partial f}{\partial \overline{Z}}\right)\,{}^t(C\overline{\Omega}+D).
\end{aligned}$$

Therefore

$$\sigma\left(\mathrm{Im}\,\tilde{\Omega}\,\frac{\partial}{\partial \tilde{Z}}{}^t\!\left(\frac{\partial}{\partial \overline{\tilde{Z}}}\right)\right)=\sigma\left(\mathrm{Im}\,\Omega\,\frac{\partial}{\partial Z}{}^t\!\left(\frac{\partial f}{\partial \overline{Z}}\right)\right).$$

By the induction on m, we can compute the Laplacian Δ_Ω. □

We let $L^2(A_\Omega)$ be the space of all functions $f : A_\Omega \longrightarrow \mathbb{C}$ such that

$$||f||_\Omega := \int_{A_\Omega} |f(Z)|^2 dv_\Omega,$$

where dv_Ω is the volume element on A_Ω normalized so that $\int_{A_\Omega} dv_\Omega = 1$. The inner product $(\ ,\ )_\Omega$ on the Hilbert space $L^2(A_\Omega)$ is given by

$$(f,g)_\Omega := \int_{A_\Omega} f(Z)\,\overline{g(Z)}\,dv_\Omega,\quad f,g\in L^2(A_\Omega). \tag{2.8.4}$$

Theorem 31. *The set* $\{\, E_{\Omega;A,B} \mid A,B\in \mathbb{Z}^{(m,n)}\,\}$ *is a complete orthonormal basis for* $L^2(A_\Omega)$*. Moreover we have the following spectral decomposition of* Δ_Ω*:*

$$L^2(A_\Omega)=\oplus_{A,B\in\mathbb{Z}^{(m,n)}}\mathbb{C}\cdot E_{\Omega;A,B}.$$

Proof. Let

$$T=\mathbb{C}^{(m,n)}/\big(\mathbb{Z}^{(m,n)}\times\mathbb{Z}^{(m,n)}\big)=\big(\mathbb{R}^{(m,n)}\times\mathbb{R}^{(m,n)}\big)/\big(\mathbb{Z}^{(m,n)}\times\mathbb{Z}^{(m,n)}\big)$$

be the torus of real dimension $2mn$. The Hilbert space $L^2(T)$ is isomorphic to the $2\,m\,n$ tensor product of $L^2(\mathbb{R}/\mathbb{Z})$, where $\mathbb{R}/\mathbb{Z}$ is the one-dimensional real torus. Since $L^2(\mathbb{R}/\mathbb{Z})=\oplus_{k\in\mathbb{Z}}\mathbb{C}\cdot e^{2\pi i k x}$, the Hilbert space $L^2(T)$ is

$$L^2(T)=\oplus_{A,B\in\mathbb{Z}^{(m,n)}}\mathbb{C}\cdot E_{A,B}(W),$$

where $W=P+i\,Q,\ P,Q\in\mathbb{R}^{(m,n)}$ and

$$E_{A,B}(W):=e^{2\pi i\,\sigma({}^tAP+{}^tBQ)},\quad A,B\in\mathbb{Z}^{(m,n)}.$$

The inner product on $L^2(T)$ is defined by

$$(f,g) := \int_0^1 \cdots \int_0^1 f(W)\,\overline{g(W)}\,dp_{11}\cdots dp_{mn}dq_{11}\cdots dq_{mn}, \tag{2.8.5}$$

where $f,g \in L^2(T),\ W = P + iQ \in T,\ P = (p_{kl})$ and $Q = (q_{kl})$. Then we see that the set

$$\left\{ E_{A,B}(W) \mid A, B \in \mathbb{Z}^{(m,n)} \right\}$$

is a complete orthonormal basis for $L^2(T)$, and each $E_{A,B}(W)$ is an eigenfunction of the standard Laplacian

$$\Delta_T = \sum_{k=1}^m \sum_{l=1}^n \left(\frac{\partial^2}{\partial p_{kl}^2} + \frac{\partial^2}{\partial q_{kl}^2} \right).$$

We define the mapping $\Phi_\Omega : T \longrightarrow A_\Omega$ by

$$\Phi_\Omega(P + iQ) = (P + QX) + i\,QY, \tag{2.8.6}$$

where $P + i\,Q \in T,\ P, Q \in \mathbb{R}^{(m,n)}$. This is well defined. We can see that Φ_Ω is a diffeomorphism and that the inverse Φ_Ω^{-1} of Φ_Ω is given by

$$\Phi_\Omega^{-1}(U + i\,V) = (U - VY^{-1}X) + i\,VY^{-1}, \tag{2.8.7}$$

where $U + i\,V \in A_\Omega,\ U, V \in \mathbb{R}^{(m,n)}$. Using (2.8.7), we can show that for $A, B \in \mathbb{Z}^{(m,n)}$, the function $E_{A,B}(W)$ on T is transformed to the function $E_{\Omega;A,B}$ on A_Ω via the diffeomorphism Φ_Ω. Using (2.8.5) and the diffeomorphism Φ_Ω, we can choose a normalized volume element dv_Ω on A_Ω and then we get the inner product on $L^2(A_\Omega)$ defined by (2.8.4). This completes the proof. □

References

1. A.N. Andrianov and G.N. Maloletkin, *Behaviour of theta series of degree n under modular substitutions*, Math. of the USSR Izvestija **9** (1975), 227-241.
2. L. Auslander and R. Tolimieri, *Abelian Harmonic Analysis, Theta Functions and Function Algebras on a Nilmanifold*, Springer-Verlag, New York/Berlin, 1975.
3. D. Bump and Y. J. Choie, *Derivatives of modular forms of negative weight*, Pure Appl. Math. Q. **2** (2006), no. 1, 111-133.
4. P. Cartier, *Quantum Mechanical Commutation Relations and Theta Functions*, Proc. of Symposia in Pure Mathematics, **9**, A.M.S., 1966, 361–383.
5. P. Deligne, *Groupe de Heisenberg et Réalité*, Jour. of American Math. Soc. **4** No.1 (1991), 197–206.
6. G. Faltings and C.-L. Chai, *Degeneration of Abelian Varieties*, EMG., band **22**, Springer-Verlag, New York/Berlin, 1990.
7. E. Freitag, *Siegelsche Modulfunktionen*, Grundlehren de mathematischen Wissenschaften **55**, Springer-Verlag, Berlin-Heidelberg-New York (1983).
8. D. Grenier, *An analogue of Siegel's ϕ-operator for automorphic forms for $GL(n, \mathbb{Z})$*, Trans. Amer. Math. Soc. **331**, No. 1 (1992), 463-477.
9. Harish-Chandra, *Representations of a semisimple Lie group on a Banach space. I.*, Trans. Amer. Math. Soc. **75** (1953), 185-243.
10. Harish-Chandra, *The characters of semisimple Lie groups,* Trans. Amer. Math. Soc. **83** (1956), 98-163.
11. S. Helgason, *Differential operators on homogeneous spaces,* Acta Math. **102** (1959), 239-299.
12. S. Helgason, *Groups and geometric analysis,* Academic Press, New York (1984).
13. R. Howe, *Perspectives on invariant theory: Schur duality, multiplicity-free actions and beyond,* The Schur lectures (1992) (Tel Aviv), Israel Math. Conf. Proceedings, **vol. 8** (1995), 1–182.
14. J. Igusa, *Theta functions*, Springer-Verlag, New York/Berlin, 1972.
15. M. Kashiwara and M. Vergne, *On the Segal-Shale-Weil Representations and Harmonic Polynomials*, Invent. Math. **44** (1978), 1–47.
16. A. A. Kirillov, *Unitary representations of nilpotent Lie groups*, Russian Math. Surveys **17** (1962), 53–104.
17. A. A. Kirillov, *Elements of the Theory of Representations*, Springer-Verlag, New York/Berlin, 1976.
18. A. A. Kirillov, *The Orbit Method, I: Geometric Quantization*, Contemporary Mathematics **145** (1993), 1–32.
19. A. W. Knapp, *Representation Theory of Semisimple Groups,* Princeton University Press, Princeton, New Jersey (1986).

20. A. Korányi and J. Wolf, *Generalized Cayley transformations of bounded symmetric domains,* Amer. J. Math. **87** (1965), 899-939.
21. G. Lion and M. Vergne, *The Weil representation, Maslov index and Theta seires*, Progress in Math., **6**, Birkhäuser, Boston, Basel and Stuttgart, 1980.
22. H. Maass, *Die Differentialgleichungen in der Theorie der Siegelschen Modulfunktionen,* Math. Ann. **126** (1953), 44–68.
23. H. Maass, *Siegel modular forms and Dirichlet series*, Lecture Notes in Math. **216**, Springer-Verlag, Berlin-Heidelberg-New York (1971).
24. G. W. Mackey, *Induced Representations of Locally Compact Groups I.*, Ann. of Math. **55** (1952), 101–139; II. **58** (1953), 193–221.
25. H. Minkowski, *Gesammelte Abhandlungen,* Chelsea, New York (1967).
26. C. Moore, *Decomposition of Unitary Representations Defined by Discrete Subgroups of Nilpotent Groups*, Annals of Math. (2) **82** (1965), 146–182.
27. D. Mumford, *Tata Lectures on Theta I*, Birkhäuser, Boston/Basel/Stuttgart, 1983.
28. D. Mumford, M. Nori and P. Norman, *Tata Lectures on Theta III,* Progress in Math. **97**, Boston-Basel-Stuttgart (1991).
29. Y. Namikawa, *Toroidal Compactification of Siegel Spaces*, Lecture Notes in Math., **812**, Springer-Verlag, New York/Berlin, 1980.
30. J. von Neumann, *Die Eindeutigkeit der Schrödinger Operatoren*, Math. Ann. **104** (1931), 570–578.
31. I. Satake, *Fock Representations and Theta Functions*, Adv. in the Theory of Riemann Surfaces, Proc. of the 1969 Stony Brook Conf., Ann. Math. Studies **66** (1971), 393–405.
32. I. Satake, *Factors of Automorphy and Fock Representations*, Advances in Mathematics **7** (1971), 83–110.
33. I. Satake, *Algebraic Structures of Symmetric Domains,* Kano Memorial Lectures 4, Iwanami Shoton, Publishers and Princeton University Press (1980).
34. G. Shimura, *Invariant differential operators on hermitian symmetric spaces*, Ann. of Math. **132** (1990), 237-272.
35. C. L. Siegel, *Symplectic Geometry,* Amer. J. Math. **65** (1943), 1-86; Academic Press, New York and London (1964); Gesammelte Abhandlungen, no. 41, vol. II, Springer-Verlag (1966), 274-359.
36. C. L. Siegel, *Gesammelte Abhandlungen I-IV*, Springer-Verlag(I-III: 1966; IV: 1979).
37. C. L. Siegel, *Topics in Complex Function Theory: Abelian Functions and Modular Functions of Several Variables*, vol. III, Wiley-Interscience, 1973.
38. M. Stone, *Linear transformations in Hilbert space, III. Operational methods and group theory,* Proc. Nat. Acad. Sci. U.S.A. **16** (1930), 172–175.
39. A. Weil, *Sur certains groupes d'operateurs unitaires*, Acta Math. **113** (1964), 143–211.
40. H. Weyl, *The classical groups: Their invariants and representations,* Princeton Univ. Press, Princeton, New Jersey, second edition (1946).
41. H. Weyl, *The theory of groups and quantum mechanics*, Dover Publications, New York, 1950.
42. J.-H. Yang, *Harmonic Analysis on the Quotient Spaces of Heisenberg Groups*, Nagoya Math. J. **123** (1991), 103–117.
43. J.-H. Yang, *Harmonic Analysis on the Quotient Spaces of Heisenberg Groups, II*, J. Number Theory **49**, No. 1 (1994), 63–72.
44. J.-H. Yang, *A decomposition theorem on differential polynomials of theta functions of high level*, Japanese Jour. of Mathematics, **22**, No. 1 (1996), 27–49.
45. J.-H. Yang, *Fock Representations of the Heisenberg Group* $H_{\mathbb{R}}^{(g,h)}$, J. Korean Math. Soc. **34**, No. 2 (1997), 345–370.
46. J.-H. Yang, *Lattice representations of Heisenberg groups*, Math. Ann. **49** (2000), 309–323.
47. J.-H. Yang, *A note on a fundamental domain for Siegel-Jacobi space*, Houston Journal of Mathematics **32** (2006), no. 3, 701–712.
48. J.-H. Yang, *Invariant metrics and Laplacians on Siegel-Jacobi space,* arXiv:math.NT/0507215 v1 or Journal of Number Theory **127** (2007), 83–102.

49. J.-H. Yang, *Invariant metrics and Laplacians on Siegel-Jacobi disk,* arXiv:math.NT/0507217 v1 or Chinese Annals of Mathematics **31B**, No. 1 (2010), 85–100.
50. J.-H. Yang, *The Siegel-Jacobi Operator*, Abh. Math. Sem. Univ. Hamburg **63** (1993), 135–146.
51. J.-H. Yang, *Vanishing theorems on Jacobi forms of higher degree*, J. Korean Math. Soc., **30**(1)(1993), 185-198.
52. J.-H. Yang, *Remarks on Jacobi forms of higher degree*, Proc. of the 1993 Workshop on Automorphic Forms and Related Topics, edited by Jin-Woo Son and Jae-Hyun Yang, the Pyungsan Institute for Mathematical Sciences, (1993), 33-58.
53. J.-H. Yang, *Singular Jacobi Forms,* Trans. Amer. Math. Soc. **347 (6)** (1995), 2041-2049.
54. J.-H. Yang, *Construction of vector valued modular forms from Jacobi forms,* Canadian J. of Math. **47 (6)** (1995), 1329-1339 or arXiv:math.NT/0612502.
55. J.-H. Yang, *Kac-Moody algebras, the monstrous moonshine, Jacobi forms and infinite products,* Proceedings of the 1995 Symposium on Number theory, geometry and related topics, the Pyungsan Institute for Mathematical Sciences (1996), 13–82 or arXiv:math.NT/0612474.
56. J.-H. Yang, *A geometrical theory of Jacobi forms of higher degree,* Proceedings of Symposium on Hodge Theory and Algebraic Geometry (edited by Tadao Oda), Sendai, Japan (1996), 125-147 *or* Kyungpook Math. J. **40 (2)** (2000), 209-237 *or* arXiv:math.NT/0602267.
57. J.-H. Yang, *Invariant Differential Operators on Siegel-Jacobi Space,* arXiv:1107.0509 v1 [math.NT] 4 July 2011.
58. C. Ziegler, *Jacobi Forms of Higher Degree*, Abh. Math. Sem. Univ. Hamburg **59** (1989), 191–224.

Professor Jae-Hyun Yang

Department of Mathematics
Inha University
Incheon 402-751
Republic of Korea

Heisenberg Groups, Theta Functions and the Weil Representation

ISBN 978-89-6105-599-4

KM KYUNG MOON SA
174, Wausan-ro Mapo-gu Seoul, Korea
Tel: (82) 2 332 2004
Fax: (82) 2 336 5193

Printed in Korea

Price: ₩ 10,000